DR. KENNETH H. COOPER'S

ANTIOXIDANT REVOLUTION

DR. KENNETH H. COOPER'S

ANTIOXIDANT REVOLUTION

KENNETH H. COOPER, M.D.

THOMAS NELSON PUBLISHERS
Nashville • Atlanta • London • Vancouver

Published in Nashville, Tennessee, by Thomas Nelson, Inc., Publishers, and distributed in Canada by Word Communications, Ltd., Richmond, British Columbia, and in the United Kingdom by Word (UK), Ltd., Milton Keynes, England.

Library of Congress Cataloging-in-Publication Data

Cooper, Kenneth H.
    [Antioxidant revolution]
    Dr. Kenneth H. Cooper's antioxidant revolution / Kenneth H. Cooper.
       p.   cm.
    Includes bibliographical references and index.
    ISBN 0-7852-8313-7
    1. Antioxidants.   2. Free radicals (Chemistry)—
Pathophysiology. I. Title. II. Title: Doctor Kenneth H. Cooper's antioxidant revolution.
RB170.C66   .1994
613.7—dc20   94-10134
                                      CIP

Printed in the United States of America
2 3 4 5 6 7 — 99 98 97 96 95

To my father, Dr. William Hardy Cooper,
who practiced dentistry for fifty years,
but encouraged vitamin-supplemented
good nutrition all of his life.

Contents

Acknowledgments

The timely and accurate preparation of this book required the contributions and cooperation of a veritable army of people. Although it is impossible to list all of the contributors, a few individuals warrant special recognition.

William Proctor, my professional literary collaborator, has worked faithfully for more than ten years in helping me with the preparation of my books. Bill's organization of the subject matter, research, and manuscript preparation are unequaled by any other contributor to this book, and his efforts are once again deeply appreciated.

For more than twenty-five years, my friend, agent, and editorial advisor, Herbert M. Katz, has helped in the preparation of all twelve of my books. Without his assistance, I doubt if the *Antioxidant Revolution* or any of my other books would have ever been published. Nancy Katz, Herb's wife and co-agent, has also worked tirelessly in helping to shape the scientific accuracy and readability of this manuscript.

Several of the members of my staff at The Cooper Aerobics Center in Dallas have also made significant contributions. Kathryn Miller, one of our experienced dietitians at The Cooper Clinic, was in charge of designing the antioxidant diet suggestions and food preparation techniques. She was ably assisted by two interns, Carolyn Murray and Sally Blocker, in doing most of the research required for the nutrition chapter. My technical assistant, Joe Head, did considerable research in the medical literature and in acquiring and organizing journal reprints.

My administrative assistant, Harriet Guthrie, coordinated and compiled many of the reprints and the research documents. In addition, as in the preparation of all my books, she helped manage my administrative and patient responsibilities so that I would have the time to work on this manuscript. Barbara Bartolomeo has been my personal transcriptionist for many years, and I appreciate the additional work she had to perform in helping to finalize this manuscript. All members of my staff were superb in their willingness to cooperate and even go above and beyond the requirements of their daily activities so that I could complete this project in a timely fashion.

For technical guidance and assistance, I again sought the assistance of my friend, colleague, and adviser, Dr. Scott Grundy, Chairman of the Department of Human Nutrition at the University of Texas Southwestern Medical School in Dallas. Dr. Grundy has always been generous with his time and erudite with his comments, which have contributed immeasurably to the scientific accuracy of several of my books, including this one. On this project he was assisted by Dr. Ishwarlal Jialal, one of the most knowledgeable people in the world on the subject of free radicals and antioxidants. Dr. John Duncan, Director of Exercise Physiology at The Cooper Institute for Aerobics Research, provided considerable technical advice and has continued to personally supervise all of the free radical and antioxidant studies done by our research team.

Finally, I would be greatly remiss if I did not thank my wonderful wife, Millie, our daughter, Berkley, and our son, Tyler, for being understanding when work interferes with so many family-related activities. Over the past twenty-five years, as they have grown accustomed to my work ethic, I have felt fortunate to have rarely heard a complaint. I know that my family members consider themselves to be part of a team that is trying to reshape the health habits of a nation. With such an attitude, there is ample motivation for all the Cooper family to persist faithfully in these demanding but extremely important projects.

Many, many thanks to all of you who have helped bring this exciting information to the American public.

Preface

It is difficult these days to scan a medical journal, read a magazine, open a newspaper, or even watch a commercial on television without seeing a reference to "antioxidants" or "free radicals."

Although this subject remains shrouded in mystery and confusion for many people, an understanding of the issues quite literally becomes a matter of life or death. "Free radicals"—or unstable oxygen molecules, also known as "reactive oxygen species"—have now been implicated in more than fifty medical problems, including various forms of cancer, heart disease, premature aging, cataracts, and even AIDS. The links to such a wide range of disease suggest that free radicals are not isolated, peripheral phenomena, but rather, are central actors in most human health problems.

Although some amount of free radical production is essential to the health and proper functioning of the human body, excess production of radicals may be damaging or even dangerous. Even though the possible relationship between free radical damage and various diseases had been postulated for many years, it was not until relatively recently that intensive research and the availability of new scientific technology drew us closer to definitive conclusions. We have learned that a variety of factors, including ultraviolet light, air pollution, cigarette smoke, and even overly vigorous, "ultra"-type exercise may result in the output of too many free radicals in your body.

Readers of my previous books know that I have always emphasized the importance of aerobic exercise as the foundation of any preventive medicine or wellness program. That basic recommendation has not changed. Yet as more research has been conducted, it has become clear that an appropriate exercise prescription becomes more complicated—especially when you take into account exposure to free radical "triggers" such as overtraining. In that regard, we will discuss two major topics: (1) the surprising value of low-intensity physical activity in producing health and longevity benefits, and (2) the need for antioxidant vitamin therapy to combat the potentially harmful effects of excess free radicals caused by too much physical activity.

Perhaps the most interesting, and I am sure the most controversial, topic in this book is the idea that too much exercise—sometimes referred to as "distress" exercise—may actually increase the risk of developing medical problems. Admittedly, much of this information is speculative and based to some extent on anecdotal information. Furthermore, I am in no way unequivocally discouraging high-intensity, prolonged activity in the healthy, fit person who is carefully listening to his or her body. But I am strongly recommending the use of antioxidants on a regular basis, regardless of the level of physical activity. It is quite true that the scientific data fall short of saying that beyond any doubt, ultra exercise activities definitely increase the risk of cancer, heart disease, and other degenerative problems. But there is certainly some solid research that points to the conclusion that antioxidants can prevent or delay the onset of many health problems, including cancer and heart disease.

Before you embark on any exercise or health improvement program, including this one, you should get medical clearance from a qualified physician. Be sure that your doctor knows exactly what dosages of antioxidants you are taking so that he or she can monitor any side effects, including interaction with other drugs. If you keep your doctor informed and follow precisely the recommendations in this book, you will find that the program is not only safe, but also highly beneficial for your health and longevity.

For more than thirty-five years I have specialized in bringing new medical concepts to the attention of the general public and providing specific programs to help give those concepts practical value. This book is just another step in my attempt to achieve my broader goal of preventive medicine for the largest number of people.

Finally, as you begin your adventure with the *Antioxidant Revolution*, I wish you good luck—and remind you that "the path to fitness is a journey, not a destination." So after you get started, don't expect that someday you will "arrive" or finally "make it." Rather, resolve to stay on this particular road to health for the rest of your life.

PART ONE

The Enemy Within: Are Free Radicals Destroying Your Health?

1

The Antioxidant Revolution

The case disturbed me deeply.

Werner Tersago, who was in his late forties, had been coming to me for checkups for more than ten years, and he always seemed to be one of my healthiest and fittest patients. He consistently scored in the "superior" category of endurance on his stress test. In fact, he did as well on that test as top athletes twenty years younger—a performance that reflected his marathon training and the fifty-plus miles per week he regularly ran. Only six months before I had last seen him, he won the master's division of an international marathon with an exceptional time of two hours, thirty-five minutes.

Werner's blood tests, including his cholesterol levels, were normal—apparently a product of his highly disciplined, low-fat, high-fiber diet. Also, his blood pressure was well within normal limits. His only physical concession to his forty-nine years was some loss of hair. All in all, Werner was a striking physical specimen.

Then it hit him—a series of headaches culminating a year later in a severe headache with some loss of equilibrium while he was skiing. The pain lingered, so he scheduled some medical tests to try to find out what was wrong. Diagnostic studies revealed a brain tumor, which at first was believed to be inoperable. Later, surgery was performed, followed by chemotherapy, which was temporarily successful. Nine months later, the second and third operations were performed, and they led to a partial paralysis.

3

Werner had to stop running, but he did learn to walk again and continued with this exercise until a few weeks before his death.

As far as anyone could tell, Werner had none of the risk factors that are usually associated with a malignant brain tumor. His diet was exemplary, and there was no history of of brain tumors among family members.

A lack of risk factors certainly does not rule out the possibility of getting a serious disease. Still, something bothered me about Werner's case, something I couldn't quite put my finger on. He had seemed almost obsessed with his running, though he had regarded it as a safety valve to control the stress in his life. After Werner's death, as I began to think more about his case, I wondered if his excessive exercise program might in any way have been a factor in his death.

As I pondered Werner's situation, another case came to mind. In 1988 my good friend Sy Mah died of cancer at sixty years of age. At the time, he held the world record for having run the most marathons—he had completed 524 races! But I wondered: Was his early death from cancer in any way related to his excessive running? Or was it linked to some other cause, such as a history of cancer among his family members?

Several similar tragedies, involving marathoners and ultramarathoners who have been diagnosed with cancer at seemingly young ages, also attracted my attention. I continued to ask, is there a correlation between excessive running and malignancies? And if so, is there something that can be done to counteract the danger?

The answers to those questions are emerging in the experiences of people like Ruth Heidrich, a highly conditioned competitive runner who seemed to be in exceptionally good health—until breast cancer struck. She had to undergo a radical mastectomy because the cancer had spread throughout her breast and possibly even into her bones and left lung.

Ordinarily, such devastating events in an otherwise vigorous forty-seven-year-old woman would cause severe depression and an overwhelming sense of futility and despair. But if you knew Ruth, you would understand that she does not respond that way. Even though she had been given less than two years to live, she decided that if she was going to die, she would die at the peak of energy and fitness. She increased the number of miles she was running—and added swimming and cycling to her regimen in preparation for the Hawaii Ironman Triathlon. But more important, she became a strict vegetarian even though she had already switched to a chicken, fish, and generally low-fat diet. The new vegetarian regimen emphasized broccoli, carrots, brown rice, and other foods that she believed would give her heavy doses of such antioxidants as vitamin C, vitamin E, and beta carotene.

Eight years later, Ruth had completed six Ironman Triathlons, more than forty marathons, and was regularly competing in middle distance races almost every weekend. Now, twelve years after her surgery, she is still in excellent health, still exercising, and still competing in demanding sports events—and she is totally cancer-free.

Ruth's experience poses two questions:

1. Did her original, highly demanding exercise program and high intake of animal foods in any way cause her cancer?
2. Did her dietary changes and increase in antioxidants slow down or stop the growth of the cancer?

As you read this book, you will see that a groundswell of scientific evidence is suggesting a resounding yes to both of those questions. Furthermore, increasing numbers of doctors who are on top of the latest medical developments are closing ranks behind patients like Ruth in advocating the use of high antioxidants, whether through the use of supplements or a very strict diet, to combat various cancers. Not only that, such physicians—and I am one of them—are increasing their own intake of antioxidants.

But the early onset of cancer in otherwise healthy and well-conditioned people has been only one of my concerns. One friend, for instance, had been a regular participant in several one-hundred-mile runs. Yet he had been forced to stop exercising due to a diagnosis of severe, obstructive three-vessel coronary artery disease.

And what about my friend Jim Fixx, author of the immensely popular 1977 book, *The Complete Book of Running*? In my book, *Running Without Fear*, I explored his history in great detail, but I have concluded that there is more to the story than I had originally thought.

At age fifty-two, Fixx died after completing a four-mile run in Vermont. The autopsy revealed that his coronary arteries, the vessels that feed blood to the heart, were almost completely blocked. There was also clear evidence of scar tissue from two previous heart attacks. In the seventeen years preceding his death, Fixx had run thirty-seven thousand miles, including twenty marathons, and he was still running sixty miles per week up to the time of his death.

Of course, it could be contended that if Fixx and other highly trained, older athletes had not been involved in marathoning and ultramarathoning, they might have succumbed to their disease many years before. In other words, as the argument goes, heavy exercise may have afforded them some protection from an even earlier death. But increasingly, I have come to believe that *there may also be a link between overtraining and disease.*

My own personal experience confirms this conclusion. Having run several marathons myself, I was always disturbed that during the final few weeks of training, I frequently came down with a viral type of illness, such as a cold or the flu. So at those very times that I needed to concentrate most intensely on my training, I was forced into a few days of inactivity.

When those bouts of sickness first hit me, I thought I had become another victim of the "cold season" or the "flu bug." But then I discovered a study done in conjunction with the 1987 Los Angeles Marathon. A questionnaire that was sent to all the participants after the race revealed that 40 percent of the runners experienced at least one cold or flu episode during the two months preceding the marathon. Even more striking, 13 percent of the

2,300 marathoners caught colds in the week immediately following the race (*Runners World*, March 1990).

Such reports have caused me to ask additional questions: Does intensive exercise weaken the immune system? And if so, why?

A comprehensive medical mystery was beginning to emerge. A cursory study of my patient and research files and reports from other physicians had demonstrated that the only common thread in those cases was intense exercise programs. In every situation, the sick people were engaging in what is sometimes referred to as "distress" exercise. Among other things, they were subjecting themselves to the chronic physical fatigue and frequent injuries that often accompany overtraining.

Finally, there was an interesting other-side-of-the-coin twist to observations linking exercise and disease. A series of studies carried out at the Cooper Institute for Aerobics Research has shown that moderate exercise can be just as effective as intense training in reducing deaths from all causes and in prolonging life.

So increasingly I wondered if it was possible that excessive exercise might not only be unnecessary, but might even be harmful— so harmful that my own physical training philosophy and recommendations should be altered. Given my previous efforts to promote the acceptance of regular exercise as a prerequisite to good health, I believed that these were landmark issues that had to be resolved.

The first great revolution in fitness, for which I have widely received credit, has unquestionably saved millions of lives. That was the introduction of aerobic exercise to the general public in 1968 through publication of my first book, *Aerobics*. Since then, the statistics are irrefutable. Mortality rates for coronary heart disease and stroke in the United States have declined by 50 and 57 percent respectively during the past two decades. Today, as a result of regular endurance exercise, better diet, reduced cigarette smoking and medications, deaths from heart disease and stroke are still declining. (See the *Joint Statement from the National*

Heart, Lung and Blood Institute and the American College of Preventive Medicine, 1994.)

Such findings have established without question the important role of physical conditioning in any program for good health. But a paradox has emerged. It is now necessary to question the claim that if something is good, more is even better. A prevalent assumption, which we now must challenge, is that the more oxygen your lungs and heart can process, the healthier you will be. In other words, is it possible to be fit but unhealthy?

Those questions have been transformed into convictions as research has emerged showing clearly that not all oxygen is good oxygen. Indeed, we now know that a number of environmental factors and seemingly beneficial health habits—including exercise to excess—can harm our health by triggering the release in the body of unstable oxygen molecules known as "free radicals."

The Enemy Within

You can't see them. You can't feel them. They leave behind only fleeting traces of their presence. But make no mistake: Your heart, your lungs, your blood vessels—all your organs and tissues—are under constant attack by wide-ranging teams of biological renegades. Even as you hold this book in your hands, no part of your body is sheltered from the destructive assaults of these molecular outlaws, which are known as "free radicals."

The bombardment that your system is sustaining every single day can be devastating. In fact, many experts believe that free radicals pose one of the greatest single threats to our public health as we approach the brave, new world of the twenty-first century. The latest research shows clearly that these lethal enemies to your health and life have solid links to the following:

- *Heart and blood vessel disease.* Free radicals seem to be the real culprits in damaging the LDL (known as the "bad" cholesterol because it has been linked to the build-up of plaque in the arteries). Unless LDL becomes damaged or "modified," it seemingly is not harmful. The damaging of LDL thus appears to be a critical link between high blood cholesterol and the build-up of vessel-blocking cholesterol plaques, called atherosclerosis. Atherosclerosis is the major cause of hardening of the arteries and heart attack. Also, free radicals may be associated with low levels of HDL cholesterol (the "good" cholesterol that has been tied to increased protection against cardiovascular disease. See chapter 6 for a more complete discussion of the different forms of cholesterol).

- *Cancer.* Radicals have been implicated in cancers of the lungs, cervix, skin, stomach, prostate, colon, and esophagus.
- *Cataracts.* Cloudiness or loss of transparency of the lens of the eye may result from the impact of free radicals.
- *Aging.* The breakdown and sagging of skin tissues and deterioration of bodily organs, which are associated with the aging process, are aggravated by free radicals. Much of the damage occurs as the radicals attack your DNA molecules and "longevity determinant genes" (LDG).

Other diseases that have been linked by medical research to the insidious operation of free radicals in your body read like the index of a medical encyclopedia. They include more than fifty conditions such as stroke, asthma, pancreatitis (inflammation of the pancreas), inflammatory bowel diseases such as diverticulitis, ulcerative colitis, peptic ulcers, chronic congestive heart failure, Parkinson's disease, sickle cell disease, leukemia, rheumatoid arthritis, bleeding within a cavity of the brain, and high blood pressure.

Whatever your health concern, you may assume that somehow, in some cloaked and devious fashion, free radicals may be a factor in causing or exacerbating the condition. It also follows that they are significantly increasing your risk of premature death. As we approach the end of this century, free radicals are indeed emerging as a great, new danger to our overall health and well-being.

In recent years, evidence has mounted regarding the benefits of regular, moderate aerobic activity, low-fat/high-fiber diets, and stress reduction programs. Now, it is time to add a second major offensive—a new revolution, if you will—to your personal health and fitness program. But you must first know your enemy before you can build a firm defense.

What Exactly Are Free Radicals— and How Do They Operate?

To understand what free radicals are and how they work, it is necessary to embark on a flight of the imagination. So let's take a brief journey from the world of the seen to the submicroscopic world of the unseen, where our health is influenced by the interaction of atoms and electrons.

Assume that you could shrink a TV camera to the size of a molecule and somehow insert it into your own body—perhaps aboard a tiny submarine that is capable of exploring your system via the many inner fluids that sustain your life. In a sense, we have a kind of *Fantastic Voyage* revisited, but with a different purpose from the one in the story. Your goal: to identify and observe the activity of free radicals.

As the camera begins to move about inside your body at subcellular levels, you immediately become aware of a process that might be described as biological fireworks. You see unstable oxygen molecules darting crazily about and crashing into other particles and tissues. Chemical studies have demonstrated that the impact of those particles actually produces bursts of light. Their movement and appearance are volatile and unpredictable in comparison with other molecules because they have one or more unpaired electrons in their outer orbits. That deficiency in their structure causes them to seek out other molecules with which they can combine. In some ways, they are like powerful internal magnets that must latch onto something else if they are to gain any semblance of stability.

Scientists have given those unstable oxygen molecules the rather adventuresome name "free radicals." Other molecules, known as "free oxygen species," behave much the same way as free radicals, though they are constructed somewhat differently. For present purposes, I will place both of these molecular loose cannons under the heading "free radicals."

Keep in mind that there are both unstable and stable oxygen molecules shooting about in your body. The stable oxygen is absolutely essential to sustain life. The point must also be made that some unstable oxygen molecules (free radicals) are good in that they enable you to fight inflammation, kill bacteria, and control the tone of your "smooth muscles," which regulate the working of your internal organs and blood vessels.

The key to the effective and safe operation of free radicals in your body is balance, but the problem is that the delicately balanced mechanisms frequently get out of whack. To correct the situation, your body produces free radical scavengers—known as "endogenous antioxidants"—which gobble up the extra free radicals and prevent them from damaging your body.

Some antioxidants that we take in from the outside through our diets help to bolster our defenses against excessive numbers of free radicals. The most important of those outside (or "exogenous") antioxidants are vitamin C, vitamin E, and beta carotene.

Unfortunately, the normal internal and external protective systems often are not adequate. The problem is that too many free radicals may be generated by such factors as air pollution, cigarette smoke, ultraviolet light produced by the sun, pesticides, and other contaminants in your food—and even too much exercise. It seems that everywhere we turn, substances and situations threaten to flood our bodies with free radicals.

When your body becomes overwhelmed by extra free radicals, those unstable oxygen molecules are transformed from your allies into molecular predators. They begin to run wild, successfully attacking healthy as well as unhealthy parts of the body. Heart disease, various cancers, and many other diseases are frequently the result.

Although we will examine this subject in more detail in later chapters, here is a brief overview of how free radicals may ravage different parts of your body and your health.

Coronary Artery Disease

This condition, which is the major cause of heart attacks, occurs when there is atherosclerosis in the vital coronary arteries, which supply blood and nutrients to the interior of the heart muscle. Excess "bad" cholesterol (LDL cholesterol) has been tabbed as the main culprit in forming atherosclerosis in the coronary arteries. But we now believe that free radicals may be the key to the development of the cholesterol-clogging atherosclerosis.

It appears that the clogging occurs after an LDL particle becomes oxidized inside the wall of a blood vessel by exposure to free radicals. The white blood cells (macrophages) in the artery walls attempt to remove the damaged LDL particles by "eating them up." Unfortunately, after ingesting the LDL, the cells can't rid themselves of the cholesterol portion of the LDL. They become engorged with cholesterol and swell up, a process that leads to a thickening of the artery wall and narrowing of the coronary arteries.

So it isn't the LDL cholesterol by itself that blocks the artery. Instead, it is the oxidized LDL, gobbled up by the white blood cell, that does the damage.

The oxidizing radicals may be activated by a variety of factors, including cigarette smoke, air pollution, and excessive exercise. At first, the LDL may be able to fend off the radical attack through an arsenal of antioxidants it has on board, such as vitamin E and vitamin C, which enhances the effect of the E. In effect, the antioxidants sacrifice themselves for the LDL. But before long, if there are too many radicals engaging in the battle, the

LDL's antioxidants are depleted. The LDL is then left defenseless.

Soon, the radicals inflict mortal wounds on numerous particles of LDL cholesterol and cause them to be consumed by the white blood cells, or macrophages. The resulting swollen cells, called foam cells, become lodged in the artery wall each time the process occurs. An accumulation of foam cells is what produces plaque, which causes a narrowing of the arteries, and eventually, serious cardiovascular disease and perhaps a heart attack.

Many heart attacks occur in the prime of life in people with no elevation of either total or "bad" LDL cholesterol. In such cases, the usual conclusion is that the heart problem was related to stress. In fact, it has been shown that stress may induce catastrophic heart rhythm irregularities which, if not promptly treated, may lead to cardiac arrest and death.

One of the major causes of stress that you can experience relates to being depressed. Bernard Lown, M.D., chairman of the Lown Cardiovascular Research Foundation, described the depression problem in this way:

> Social isolation, alienation from people, bereavement, being in a no-exit and stressful work situation, loss of job, lack of education, and poverty are all substantial risk factors for cardiovascular disease. On the other hand, people who have tight social networks, such as religious affiliations or communal activities, or close extended family ties have less coronary artery disease.

Since it is well known that stress increases free radical production, perhaps the presence of the molecular outlaws in the midst of depression or other emotional difficulties explains why a heart attack may occur in a person without heart symptoms and without any of the coronary risk factors, such as cigarette smoking, high cholesterol, elevated blood pressure, or obesity.

Cancers

Free radicals operate in a somewhat different way with cancers. They are still triggered by such environmental factors as cigarette smoke, pollution, and ultraviolet radiation, and possibly in response to stress and overtraining. But in the case of cancers, the radicals shoot deep into the inside of nearby cells and damage the nucleus, which carries DNA, the genetic code of the cell.

As a consequence, the cell may grow out of control, with malignant lesions and tumors being the final result. When the radicals strike a cell, they tend to start a chain reaction of other radicals, which shoot out toward other cells, so the potential for damage is multiplied. The best line of defense is to preserve and build up the antioxidant shields in your body—a defense that centers on a new concept of exercise, diet, and supplement programs explained in this book.

The Membrane Wars

Some free radicals may do most of their damage to the various membranes and tissue coverings in your body. For example, they may injure the lenses of your eyes and cause cataracts, or they may attack skin tissues and foster premature aging. Yet women and men have reported to me that a few months after starting an antioxidant program such as the one described in this book, they have noticed smoother and more pliable skin and the disappearance of dry, cracked areas on the elbows and other parts of the body.

As you view your imaginary TV screen and watch the destruction being wrought by these biological outlaws, the first thought that may come to mind is, *My body isn't doing the job. My protective shields just aren't working properly. What steps can I take to protect myself?*

The answer: Join the Antioxidant Revolution.

Joining the Revolution

As I have already indicated, the Antioxidant Revolution is really the second major transformation in personal health that I have promoted. In some respects, this revolution builds upon the first one, which centered on aerobic exercise, a low-fat diet, and other non-drug approaches to good health. But the Antioxidant Revolution also marks a shift in my thinking about exercise and physical fitness.

The first revolution began more than twenty years ago when I started the Cooper Clinic and Cooper Institute for Aerobics Research in Dallas. At that time, many members of the medical establishment questioned my motives. They thought it was bizarre and unprofessional to promote exercise and the gospel of preventive medicine.

But since I first set up shop in those early days, a tremendous change has occurred in health care. Increasingly both doctors and patients have begun to understand that medicine and surgery alone cannot overcome our worst diseases. They have finally started to accept the fact that changes in lifestyle, such as regular aerobic exercise and a low-fat, high-fiber diet, are fundamental to good health and long life. And it is cheaper and more effective to maintain good health than it is to regain it once it is lost.

It is evident, then, that the first revolution has been won. But what about the second?

The Antioxidant Revolution will take us another step beyond my first book, *Aerobics,* and my 1988–1989 best-seller, *Controlling Cholesterol.* In formulating the new programs in this book, I have relied on several internationally recognized consultants and experts, including Scott M. Grundy, M.D., Ph.D., the director of the Center of Human Nutrition, University of Texas Science Health Center at Dallas. In addition, the Cooper Institute for Aerobics Research and my alma mater, the Harvard School of Public Health, have been involved in ongoing research which has a direct bearing on the question of free radicals, antioxidants, and disease.

The Cooper Institute has also sponsored a special research project on free radicals, which has been coordinated by Ishwarlal Jialal, M.D., director of clinical chemistry, the University of Texas

Southwestern Medical School in Dallas. Among other things, this investigation has explored the extent to which trained and untrained individuals generate destructive free radicals during exhaustive exercise—and how antioxidants can help counter the danger.

As a result of those studies, the Antioxidant Revolution will provide you with many new insights and programs to help you protect yourself from the free radical threat. You may be surprised to learn that:

- The amount of the antioxidant vitamins you should take—regardless of your sex, age, or activity level—is higher than you have been led to believe in many popular media accounts.
- All adult men require more of some antioxidant supplements than adult women.
- Men and women more than fifty years of age need more of certain antioxidants than younger males or females of all ages.
- Boys and girls who are heavy exercisers need more antioxidants than less-active adults.
- Some people—and you may be one—should avoid certain antioxidants altogether because of adverse side effects.
- You should limit the amount of certain exercises—but increase other types in order to bolster your internal protection against disease. To assist you, I have included a modified fitness point system, based on new findings about the production of free radicals during exercise. You will also learn at what level of strenuous exercise you should consume extra antioxidant supplements, so as to prevent injury to your body.
- It is possible to reduce soreness and fatigue after intense physical activity by employing certain little-known but scientifically proven principles about antioxidants.
- Simple principles of food preparation, based on the latest

research into nutrition, can greatly reduce your loss of dietary antioxidants—a loss that most people experience in normal cooking and eating.

- It is essential to manage your own environment so that you minimize your exposure to the forces that trigger the release of free radicals in your body.

Now the time has arrived for you to begin your own personal Antioxidant Revolution. Our first objective is to explore more thoroughly how those tiny molecular outlaws, the free radicals and their kin, have burst forth on the scientific scene—and how they may be working in different parts of your body.

2

Unmasking the Free Radical Threat: The Latest News from the Medical Front

The focus on the impact of free radicals and antioxidants on your health is a relatively recent phenomenon. Until the beginning of this century, no one knew that free radicals could exist and operate independently. In fact it has been only in the last three to four decades that our scientific understanding of the free radical/antioxidant connection has emerged.

Increasingly this new knowledge is being applied by practicing physicians in examining rooms and clinics around the country, but much in the scientific literature still has not filtered down to the average patient. So to be able to discuss important issues with your doctor, and to ask incisive questions, it is essential that you be aware of events on the medical research front. Only with this information can you hope to understand the full range of what free radicals can do *to* you—and what antioxidants can do *for* you.

I have identified five major breakthroughs in this turbulent field of scientific investigation.

1. Unmasking the true nature of free radicals;
2. Revealing the dark side of oxygen;
3. Uncovering the body's natural defenses against free radicals;
4. Pinpointing free radical cell damage as a probable cause of cancer—and offering antioxidant remedies; and
5. Linking oxidation to atherosclerosis.

Now let me introduce you to each of these breakthroughs and to the scientists who have been the driving force behind them. (For a more detailed scientific explanation of the basis of my Antioxidant Revolution program, see appendix 2.)

Breakthrough #1:
Unmasking the Free Radicals

Chemists in the nineteenth century used the term "free radical" to refer to a group of atoms that form a molecule. In those days, scientists did not believe that free radicals could exist independently, or in a free state.

But things changed dramatically with the advent of the twentieth century and the work of Russian expatriate Moses Gomberg. He prepared the first independent organic free radical, triphenylmethyl, in a yellowish solution in his laboratory at the University of Michigan in 1900. In performing that experiment, he derived the free radical from triphenylemethane, a hydrocarbon that serves as the basis for many dyes.

Born in Blisavetgrad, Russia, in 1866, Professor Gomberg emigrated with his family to the United States at the age of eighteen. His father's anti-czarist activities had made the family *persona non grata* in Russia, and they had to flee with little more than the clothes they were wearing. In a classic American success story, Gomberg rose above his poverty and lack of knowledge of English to receive a doctorate at the University of Michigan in 1894. His family's radical political background was mirrored in his focus on molecular renegades in the laboratory, and Gomberg went on to become the head of the chemistry department at Michigan in 1927 and remained there until his retirement in 1936.

As a result of the research of Gomberg and other scientists during the first part of the twentieth century, the term "free radical" came to mean a relatively unstable molecule with one or more unpaired electrons. As those single electrons move about in their orbit in the molecule, they create a kind of magnet effect, which causes the free radical to combine with nearby molecules.

Many free radicals are so unstable that they can exist for only a fleeting moment, a microsecond. During their brief life, the radicals act as catalysts, or bridges, to spark chemical reactions and changes in other molecules. The high-speed, highly interactive quality of many free radicals was identified in experiments by Friedrich Adolf Paneth, an Austrian chemist, who collaborated with W. Hofeditz, a German researcher. In 1929, they discovered the brief and powerful existence of the methyl and then the ethyl free radicals.

Paneth's life was also a metaphor for his work in the laboratory. He seems to have had as many political problems as Gomberg— and might be considered something of a radical in his society. As the Nazi movement grew in Central Europe before the outbreak of World War II, he headed for England, where he eventually became professor of chemistry at the University of Durham. He later returned to West Germany as director of the Max Planck Institute at Mainz and pursued landmark studies in rare gases and the composition of the atmosphere.

Breakthrough #2:
Revealing the Dark Side of Oxygen

It was not until 1954 that the destructive power of free radicals on living organisms, including the human body, was recognized. Ironically, the culprit that received the finger of blame was the main support of life on earth, oxygen. Scientists during the 1940s and early 1950s had cataloged in their experiments a host of mysterious injuries to biological tissues. Fish, rats, and other animals suffered tissue damage, lower growth rates, and other injuries when they were exposed to high concentrations of oxygen. In humans, breathing pure oxygen for as short a period as six hours caused chest soreness, coughing, and sore throats—and longer periods of exposure could destroy the air cells in the lungs.

In addition, a form of blindness known as retrolental fibroplasia appeared among premature infants. The disease, which involved the formation of fibrous tissue behind the lens of the eyes, became widespread during the 1940s—and the medical establishment became increasingly puzzled. Finally, in 1954, scientific sleuths determined that the source of the problem was incubators, where the premature newborns were placed in an atmosphere with much higher amounts of oxygen than the 21 percent contained in ordinary air. Two of the sleuths, American scientists Rebecca Gershman and Daniel L. Gilbert, linked the development of retrolental fibroplasia in premature babies to oxygen free radicals. In fact, they concluded that most of the damage done to living tissues was the result of oxygen radicals.

Even Joseph Priestly, the English chemist and cleric who discovered oxygen in 1774, questioned whether the gas, which is so essential to life, might also in some way be harmful. But he lacked scientific training and also was hamstrung by the fact that experimental chemistry was in its infancy in his day. Priestly's suspicions remained in the realm of speculation until the middle of the twentieth century, when Gershman, Gilbert, and other researchers began to nail down our present understanding of the true dangers to human health of certain outlaw forms of oxygen.

In the years that have followed the 1954 breakthrough, four extremely destructive forms of oxygen have been identified. Two of them—the hydroxyl radical and the superoxide radical—are true free radicals in that they have an unpaired electron in a molecular orbit. Two other renegade forms of the oxygen molecule, known as "non-radical reactive oxygen species," can also do significant damage to the body. Labeled the oxygen singlet and hydrogen peroxide, these forms of oxygen, along with the two free radicals, are the main enemies we will be fighting through the various practical programs described in this book.

Breakthrough #3:
Discovering the Body's Defenses
Against Free Radicals

The next major milestone in antioxidant and free radical research was achieved in 1968, when American scientists J. M. McCord and I. Fridovich discovered a natural antioxidant enzyme in the human body, superoxide dismutase (now commonly referred to as "SOD"). A popular assumption today is that most antioxidants enter the body from the outside through the diet or dietary supplements in the form of vitamins E, C, and beta carotene. But the 1968 breakthrough established that the body also possesses an important inner "police force" in the form of "endogenous," or internally produced, antioxidants.

McCord and Fridovich determined that the primary purpose of SOD is to remove the destructive free radical superoxide—one of the four major molecular outlaws mentioned above. Shortly afterward, they developed a landmark concept know as the "superoxide theory of oxygen toxicity." The theory states that the superoxide radical is a major cause of the damage inflicted by unstable oxygen in the body, and that SOD is the body's primary defense.

It was only a short step from the formulation of that theory to the implication of other free oxygen radicals in the body's tendency to deteriorate over time and to be victimized by various diseases. Fridovich, the discoverer of SOD, was also the first scientist to demonstrate that the most powerful and destructive free radical known to science, the oxygen renegade hydroxyl radical, can be formed in a biological system such as the human body. Increasingly, the evidence pointed toward a link between free radicals like the hydroxyl radical and cancer.

Breakthrough #4:
Identifying Free Radical Cell Damage
as a Cause of Cancer—and Antioxidants
as the Answer

During the 1970s, scientists such as Lester Packer, a biochemist at the University of California at Berkeley, began to hone in on free radicals as a major threat to human health. At the same time, Packer has noted in a conversation with *The New York Times Magazine* that "we need free radicals to live"—for several reasons.

For one thing, Packer and others have determined that free radicals benefit the body by working with the immune system to ward off disease by killing alien bacteria and other invaders that enter your body. Also, they help regulate the contraction of the smooth muscles of your blood vessels and contribute to the control of your blood flow by influencing the tone of the tissue lining of your vessels.

Free radicals are released during the normal metabolism of your body, as your food is turned into energy by your body's cells. Your body's defense systems, including the antioxidant enzymes SOD, catalase, and GSH, exist to keep the output of free radicals in balance. The problem arises when too many free radicals are generated for your internal antioxidant police force. When that happens, the radicals become renegades.

"The hydroxyl radical is a very destructive character," said Professor Packer, referring to the outlaw oxygen molecule that has been called the most reactive radical known to chemistry.

That radical, as well as a number of other unstable oxygen molecules, has been linked to many serious diseases, including cancer. The primary means by which radicals cause cancer is by launching an attack on the nucleus of cells and damaging the DNA, which may lead to cell mutations.

"Research data indicates that reactive oxygen species are involved in the process of cancer initiation and promotion," Packer said in 1991. "Increased incidence of cancer with advancing age may be due, at least in part, to the increasing level of free-radical reactions with age, along with the diminishing ability of the immune system to eliminate the altered cells."

Researchers reporting in the *Journal of the American Medical Association* in 1994 explored whether aging of the population and smoking patterns completely accounted for increased rates of cancer mortality and cancer incidence among whites from 1973 through 1987. They found that while cigarette smoking is the single most important known cause of cancer and other chronic diseases today, about 70 percent of cancer *isn't* generally linked to smoking. (See the Feb. 19, 1994 issue of JAMA, Vol. 271, No. 6.) Among the other factors that probably are contributing to the increased incidence of cancer, a prime candidate must be excessive free radical exposure due to the pollution of our environment.

A link may also exist between free radical damage and stress. Stressful life events, especially job-related problems, may greatly increase the risk of cancer of the colon and rectum, according to a study published in *Epidemiology* in July 1993. In evaluating more than one thousand Swedes, researchers compared medical histories and questionnaires, which focused on important life events during a ten-year period. They found that serious work-related problems made a person five times more likely to develop colo-rectal cancer and that unemployment of more than six months doubled the cancer risk. In addition, those who moved more than 120 miles had nearly three times the risk of cancer. Divorce or a spouse's death increased cancer risk by 50 percent. The exact reasons for the apparent relationship between stress and cancer are unknown, but the increased free radical production that can result from stress may be the answer.

Fortunately, we have protective shields available in the form of antioxidants—including vitamin supplements. In this regard Packer advised, "From cell culture and animal research, it appears that vitamin E and other antioxidants alter cancer incidence and growth through their action as anticarcinogens, quenching free radicals or reacting with their product." He emphasized that epidemiological human studies, though limited, suggest that "vitamin E and the other antioxidants may decrease cancer incidence."

Professor Lester Packer, along with a pantheon of other scientific experts, has helped build an indisputable fortress of research findings, which clarify the connection between free radicals and cancer—and the power of antioxidants to counter that threat.

Among other things, these investigators have discovered a link between lung cancer and low beta carotene. Low intake of fruits and vegetables, especially those with yellow-reddish or orange pigmentation such as carrots, is consistently associated with increased risk of lung cancer, according to Regina G. Ziegler of the National Cancer Institute.

"In addition, low levels of beta-carotene in serum or plasma are consistently associated with the subsequent development of lung cancer," Ziegler said. "The simplest explanation is that beta-carotene is protective."

A 1994 study conducted by the National Cancer Institute and the National Public Health Institute of Finland failed to show any reduction in the incidence of lung cancer among long-time male smokers after five to eight years of dietary supplementation with vitamin E or beta carotene. But this investigation, published in April 1994 by *The New England Journal of Medicine*, did report 34 percent fewer prostate cancers among men taking vitamin E; 16 percent fewer colon and rectal cancers among those on vitamin E; and 5 percent fewer deaths from ischemic heart disease among those smokers on vitamin E.

Other researchers have also found a connection between low beta carotene and vitamin E and various cancers. George Comstock and associates from the department of epidemiology, Johns Hopkins School of Hygiene and Public Health, collected blood specimens from nearly twenty-six thousand subjects, mostly women aged thirty-five to sixty-four, in Washington County, Maryland. They tracked the health of the participants from 1974 until 1989 and compared the blood samples with cancers that had developed.

The study established a strong association between higher levels of beta carotene in the blood and protection against lung cancer. There were "suggestive" links of higher levels of that vitamin with a lower incidence of melanoma, bladder cancer, and rectal cancer, according to a 1991 article in the *American Journal of Clinical Nutrition*. Also, higher serum vitamin E levels were associated with protection against lung cancer.

The Basel Study, which was started in Switzerland in 1959 and began to focus on antioxidants in the early 1970s, found that participants with low blood levels of beta carotene had a significantly elevated risk of lung cancer. Also risk of any cancer was higher if both beta carotene and retinol were low. Speaking for vitamin C, Gladys Block of the Division of Cancer Prevention and Control, National Cancer Institute, said in a landmark 1991 report:

> For cancers of the esophagus, larynx, oral cavity, and pancreas, evidence for a protective effect of vitamin C or some component in fruit is strong and consistent. For cancers of the stomach, rectum, breast, and cervix there is also strong evidence. . . . It is likely that ascorbic acid, carotenoids, and other factors in fruits and vegetables act jointly. Increased consumption of fruits and vegetables in general should be encouraged.

Important studies that further solidify the connections between cancer, free radicals, and antioxidants continue to proliferate. The response to this blizzard of scientific support seems clear: A sound, comprehensive antioxidant program is necessary to achieve the best chance for continuing good health.

Breakthrough #5:
Linking Oxidation and Atherosclerosis

In the early 1980s, Dr. Daniel Steinberg, professor of medicine at the University of California in San Diego, proposed the theory that oxidation—or the combining of oxygen free radicals with other LDL particles in the blood stream and tissues—is the basis for the formation of plaque and clogging of the body's vessels.

You will recall from the description in chapter 1 that free radicals shoot out from white blood cells, or macrophages, to attack the LDL, which eventually combines with the macrophages to form foam cells. That process of oxidation is what causes food to spoil or become rancid when left on the counter, exposed to the oxygen in air.

Steinberg's insight was a culminating breakthrough in a long line of scientific findings, which began in 1910, when cholesterol was found in atherosclerotic plaques. Later, in 1952, scientists discovered that oxidized lipids (or fats) were present in the plaque. Then, in 1961, researchers discovered that special types of white blood cells, macrophages, were a major component of plaque. Ten years later, in 1971, foam cells were identified in the plaque.

Steinberg put those and other studies together to develop his theory that oxidation of LDL cholesterol is the major factor in atherosclerosis and coronary artery disease. That position has been increasingly accepted by leaders in the field as further studies have demonstrated in more detail the links among free radicals, atherosclerosis, and antioxidants that can combat the disease.

In a workshop chaired by Dr. Steinberg for the National Heart, Lung, and Blood Institute in 1991, experts from around the world concluded that a major clinical trial should be authorized to investigate more fully the impact of natural antioxidants in combating atherosclerosis.

In making their recommendations, the participants noted that a strong case had already been made for the protective effect of various antioxidants, including beta carotene, which had already been linked to a significant reduction in cardiovascular diseases as a result of the Harvard Physician's Health Study. But they expected that further studies would "'fine-tune' our understanding of these natural antioxidants—beta-carotene, vitamin E, and vitamin C."

Those five breakthroughs—and the outstanding scientists who sparked them—are only the beginning of the research foundation that has been laid for the Antioxidant Revolution. My own "conversion" to the antioxidant way of thinking has involved another international array of findings, as you will see in the following chapter.

3

Designing Your Personal Defense Against the Molecular Outlaws

For more than a decade I have been following intently the various scientific developments that have identified the free radical threat and established antioxidants as the body's essential defense. But it has been only in the last couple of years that those studies have reached a kind of critical mass—a weight of authority that demands a response from our medical leadership.

As a spokesman for preventive medicine in the United States and abroad, who has sometimes gone out on a limb in the past to advocate seemingly controversial practices—such as aerobic exercise and the control of cholesterol—I do not hesitate to put my reputation on the line now for the Antioxidant Revolution. The following sampling of recent studies and reports from all parts of the globe will give you an idea of why I have come to feel so strongly about this topic—and why I have encouraged the men and women who are my patients to take practical steps to bolster their antioxidant defenses.

In China: A research team of American and Chinese scientists studied the impact of vitamin and mineral supplements on about thirty thousand male and female residents of Henan Province in north central China, where cancer death rates are among the highest in the world. During a five-year period, the investigators gave the Chinese participants daily doses of either (1) various combinations of beta carotene, vitamin E, and the mineral selenium, or (2) a placebo, or dummy pill.

Their findings, reported in 1993 in *The Journal of the National Cancer Institute,* showed that those on the supplements experienced a decrease of 13 percent in cancer death rate and a 9 percent drop in their risk of death from all causes. Most of the benefits were enjoyed by those taking a combination of beta carotene, vitamin E, and selenium. This group also had a 10 percent decrease in the rate of deaths by stroke.

Other results in this study were that deaths from stomach cancer declined by 21 percent, and those from esophageal cancer dropped by 4 percent in the group taking beta carotene supplements.

In Canada: Antioxidant supplements apparently help strengthen the human immune system by lowering the production of those chemical substances in the blood that suppress the immune functions, according to a Canadian study.

Researchers in Canada, reporting in the November 1992 issue of the British medical journal *Lancet,* gave elderly subjects small doses of vitamin E, beta carotene, and some other vitamins and minerals for a one-year period. Those receiving the antioxidants suffered from half as many colds, flu outbreaks, and other infectious diseases in comparison with a control group given a placebo pill.

Furthermore, those vitamin-taking patients who did suffer viral diseases recovered from their illnesses twice as fast as those in the control group who did not take vitamins.

In Scotland: The Scottish Heart Health Study, conducted among forty- to fifty-nine-year-old patients of 260 general practitioners throughout Scotland in 1989–1991, has confirmed the importance of a diet high in antioxidants as a defense against atherosclerosis.

Men without diagnosed heart disease who consumed the most beta carotene, vitamin C, vitamin E, and high fiber had a significantly lower risk of developing heart disease than those with the lowest intake of such vitamins. Although the women in this particular study did not show quite as much protection as in some other vitamin investigations, those with high intakes of fiber, which is often linked to an increased dietary intake of antioxidants, did have a lower risk of heart disease.

Researcher C. Bolton-Smith and his colleagues concluded in 1992 in the *European Journal of Clinical Nutrition* that the "results suggest that high dietary intake of the antioxidant vitamins may reduce risk of CHD [coronary heart disease], particularly in men, and that fibre may be equally cardio-protective in both sexes."

In Bethesda, Maryland: A National Heart, Lung, and Blood Institute Workshop evaluated the preliminary results of The Harvard Physicians' Health Study that focused on the risks of cardiovascular disease in 333 men who had experienced chronic angina (chest pains) or had undergone heart bypass operations. But the men had no prior history of heart attacks, strokes, or transient ischemic attacks (TIAs, which are a kind of "mini-stroke").

The men in the group who received beta carotene supplements enjoyed a "significant 50 percent reduction in . . . myocardial infarction (heart attacks), revascularization (reestablishment of blood supply to a part of the body, as with a bypass operation), stroke, or coronary death," declared Dr. Daniel Steinberg in a 1992 report.

In Boston, Massachusetts: Among middle-aged women, the use of vitamin E supplements has been associated with a reduced risk of coronary heart disease, according to a 1993 report from the Harvard Medical School and the Harvard School of Public Health.

Dr. Meir J. Stampfer and colleagues reported in the May 20, 1993 issue of *The New England Journal of Medicine* that more than eighty-seven thousand female nurses, aged thirty-four to fifty-nine, who were free of diagnosed cardiovascular disease and cancer, were selected in 1980 to participate in the study. After the health of the women was followed for eight years, the investigators found that women in the top fifth in consumption of vitamin E in their diets had a significantly lower risk of coronary disease than those in the bottom fifth.

The women who had the lower risk of coronary artery disease took daily supplements of 100 IU of vitamin E over a period of more than two years. (Vitamin E is usually sold in "international units," or "IU." One international unit is equal to approximately one milligram of the vitamin.)

In a related study, reported in the same issue of *The New England Journal of Medicine*, researchers also found a link between high intake of vitamin E and a low risk of coronary heart disease in men. As with the women, the men in the lower-risk group took daily supplements of 100 IU of vitamin E.

In Baltimore, Maryland: A survey undertaken at the University of Maryland School of Medicine concentrated on interpreting the latest research on antioxidants and cataracts. The scientists concluded that the presence of oxygen radicals in the eye may constitute a significant risk factor in the formation of cataracts (a clouding of the lens). Radicals and other reactive oxygen species may be caused by exposing the eye to ultraviolet sunlight.

But survey coordinator Shambhu D. Varma, who reported his findings in 1991 to the American Society of Clinical Nutrition, also concluded that the damage done by the radicals "can be thwarted by nutritional and metabolic antioxidants such as ascorbate (vitamin C) and Vitamin E."

In Finland: A 1992 study of Finnish men and women who were forty to eighty-three years of age revealed that an increased risk of cataracts is associated with low blood levels of vitamin E and beta carotene.

In describing this study, the August 1993 issue of the *Mayo Clinic Health Letter* noted that scientists suspect cataracts develop in part from oxidation of the proteins in the lens of the eye. Vitamin C, vitamin E, and beta carotene are thought to prevent such clouding.

In New York City: Between 1979 and 1986, researcher Stanley Fahn of the Neurological Institute of New York and the Columbia University College of Physicians and Surgeons began to administer high doses of vitamin E and vitamin C to patients with early Parkinson's disease. Specifically, patients took 3,200 units of vitamin E and 3,000 milligrams of vitamin C daily. (Vitamin C is usually sold in milligrams (mg), and all references in this book are to milligrams.) Another group of patients who were placed under observation took no antioxidants.

At the end of the study, the onset of the full-blown disease, which was signaled by the need to take the medication Levodopa, was delayed for 2.5 years in the patients on the antioxidants.

"The results of this pilot study suggest that the progression of Parkinson's disease may be slowed by the administration of these antioxidants," Fahn concluded in a 1991 supplement to the *American Journal of Clinical Nutrition.* (Although subsequent studies have not duplicated those findings, research continues in this area.)

At the Johns Hopkins University Department of Epidemiology, in Maryland: In a 1992 survey of the impact of antioxidants on various cancers, researcher George W. Comstock and two colleagues found that high levels of beta carotene in the blood were associated with a strikingly low incidence of lung cancer.

In his report, Comstock said, "Low levels of beta carotene were most likely to be associated with subsequent cancer, but there were marked differences by cancer site."

In Finland: Eighteen lung cancer patients, all with a history of smoking, were given antioxidants in addition to their regular treatments. Among the oral supplements they took daily were 10,000 to 20,000 IU of beta carotene; 2,000 to 5,000 mg of vitamin C; and 300 to 800 IU of vitamin E.

Researcher K. Jaakkola and colleagues reported in *Anticancer Research* in 1992 that the antioxidant treatment, in combination with chemotherapy and irradiation, prolonged the survival time of patients with small cell lung cancer, as compared with other patients in similar circumstances. Also, the patients who survived longest started their antioxidant treatments earlier than those who died. The researchers noticed that the patients on antioxidants tolerated chemotherapy and radiation particularly well.

At the Gerontology Research Center, National Institute on Aging, Baltimore, Maryland: Recent research is establishing that the more damage that occurs to the body from free radicals, the shorter the life span. Studies have compared species of mammals that have relatively high rates of oxidative damage, such as mice, with those that are exposed to less oxidative damage, such as chimpanzees and humans. The consistent finding is that the more the body is exposed to free radicals, the shorter the life span will be.

Furthermore, aging expert Richard G. Cutler of the Gerontology Research Center concluded in 1991 in an *American Journal of Clinical Nutrition* supplement that the more antioxidants found in the body, the longer the individual's life span will be. In the bodily tissues, high levels of antioxidants that have been linked to a longer life include natural vitamin E, the carotenoids (such as beta carotene), and various internal enzymes like SOD. Also, longer-living humans have been shown to have higher amounts of vitamin C in their brain tissues.

But what do those studies mean to me as a practicing physician and to you as a patient who asks, "So what should I do to make the research findings a part of my life?"

From Professional Research to Practical Benefits

The message conveyed by such studies has been decisive in shaping my thinking as a physician and public health spokesman—and in influencing my own personal health practices. Some health care professionals would say, "Avoid vitamin and mineral supplements—you can get all the antioxidants you need from your diet." I disagree. In fact, I take a daily "antioxidant cocktail" myself and recommend that my patients do the same.

Others may say, "If you are healthy and have proper medical clearance, exercise all you like—it can't hurt." Again, I disagree. Unless you are a competitive athlete, you should concentrate on lower-intensity programs of the type described in this book. Otherwise, you may subject your body to damage from the production of excess free radicals.

These and other conclusions, which are firmly rooted in the latest scientific findings, have led me to formulate the Antioxidant Revolution program, which has the potential to provide you with health benefits such as the following.

Benefit #1: Increased protection from many forms of cancer.

Benefit #2: Stronger defenses against cardiovascular disease, such as atherosclerosis, heart attacks, and strokes.

Benefit #3: The preservation of your eyesight through the prevention of cataracts.

Benefit #4: A delay in the onset of premature aging.

Benefit #5: A more powerful immune system.

Benefit #6: A decreased risk of early Parkinson's disease— and a host of other major advantages for your health.

One of the most interesting examples of a person who strengthened his antioxidant defenses without even understanding fully the mechanism that was at work is a case that a colleague brought to my attention involving a man whom I will call Jeff, a seventy-nine-year-old resident of Florida who retired to the Sunshine State about eight years ago.

The Amazing Case of Jeff, the Florida Retiree

After reading arguments being made by pro-vitamin advocates back in the 1950s, Jeff began to take relatively large daily doses of vitamin E and vitamin C. Specifically, he took supplements amounting to 800 IU of E and 1,500 mg of C. In those days, taking vitamin supplements in any amounts was considered unnecessary and foolish by most physicians.

When Jeff began his regimen, he did not realize that his blood lipids were astronomically high. It was not until the mid-1970s—when he was about sixty-five years old—that he discovered through blood tests that his triglycerides were consistently over 1,000 mg/dl (they should have been 125 or lower). Also, he learned that his total cholesterol was over 400 mg/dl (that reading should have been 200 or lower).

Despite his high level of blood fats, however, Jeff had not experienced any problems with his heart or blood vessels up to that point. But his physician, quite correctly, placed him on an early cholesterol-lowering medication, which was eventually switched to Lopid and then Mevacor.

Unfortunately, the medication did Jeff little good. His triglycerides remained at very high levels, and his cholesterol could not be reduced below 300. Another unfortunate circumstance was that Jeff decided to eliminate his daily vitamin supplements in the early 1980s. Among other things, he was getting medical pressure to stop the program because it was "useless" and "quackery."

Soon afterward some problems began to occur. Jeff developed intermittent claudication, or pain in walking as a result of clogging of the vessels in his legs, and had to undergo surgical procedures to increase the flow of blood. For about five years, Jeff continued with his prescription medication program, although it was not working well, and he avoided antioxidant supplements. During that period he experienced two transient ischemic attacks (TIAs), or "mini-strokes," that resulted from a cut off of the flow of blood to his brain. Most likely, the TIA events resulted from the partial clogging of his carotid (neck) arteries with fatty deposits.

Finally, in 1991, Jeff began to hear about the many benefits of antioxidant therapy—and especially about the advantages of taking vitamins E and C in relatively large quantities. He learned, as we have already discussed, that vitamin E in LDL cholesterol—which is enhanced by vitamin C—provides our bodies with a major defense system to protect us from the oxidation of LDL and the formation of dangerous "foam cells." Foam cells contribute to the formation of plaque on our vessels, which leads to atherosclerosis, heart attacks, and strokes.

So Jeff returned to his practice of taking daily antioxidant supplements. A stress test, which involved having thallium injected into his blood so that the doctor could evaluate vessel blockage, was scheduled about one month before the writing of this chapter. The test revealed no significant blockage of his coronary arteries, which feed blood to the heart. In fact, Jeff's doctor told him that, contrary to all expectations considering his high levels of blood lipids, Jeff possesses a "happy heart."

Clearly, Jeff still has blood vessel problems. The vessels in his legs are clogged with plaque—though there has been no discernible progression of the disease in the last couple of years. Also, the carotid arteries leading to his brain—which became a problem after he went off his antioxidant program around 1980—remain a concern.

I find myself wondering if the clogging he has experienced might have been caused or hastened by his decision to stop the antioxidants back in 1980. In any case, for someone with such incredibly high levels of blood fats for so many years, Jeff has enjoyed an amazingly healthy and active life. And it may be that much of his good fortune can be traced to his decision back in the 1950s to take relatively large doses of vitamins E and C.

For some reason, his unusually high levels of LDL cholesterol have for the most part failed to become foam cells, which would stick to his vessel walls. The best explanation seems to be that the vitamins E and C that he was ingesting provided an effective defense for much of his life to fight off the attacks of free radicals and to prevent the formation of foam cells and plaque.

The most important lesson we can learn from Jeff's case is that antioxidants can be working to benefit your system, *even if you do not notice any changes in your cholesterol profile.* Regardless of your blood test results, it is still quite likely that you are lowering your risk of heart disease and other health problems. Take a moment to recall the reasons.

Even if your total cholesterol and "bad" LDL cholesterol stay at high levels, even if your "good" HDL cholesterol fails to rise, and even if your ratio of total cholesterol to HDL cholesterol fails to decrease, you can still lower your risk of atherosclerosis by combating the oxidation process inside your body.

Furthermore, the stronger your antioxidant defense, the less likely it is that you will develop various forms of cancer, cataracts, and a host of other diseases—and the less likely it is that you will be a victim of premature aging.

So how can you design your own antioxidant defense system and win the battle against those molecular outlaws, the free radicals? Here is a game plan for using the remaining chapters of this book to put together your own Antioxidant Revolution.

Your Antioxidant Game Plan

As you move through subsequent chapters, you will see that you need to take four steps to wage an effective fight against the dangerous free radicals in your body. At least one entire chapter will be devoted to describing each of these steps. Here is an overview and some suggestions about how you can prepare to fit the information in upcoming chapters into your own personal antioxidant game plan.

Step 1: Use the Power of Lower-intensity Exercise

In chapter 4, you will be introduced to a comprehensive set of lower-intensity exercise programs that will help you build up your endurance, or aerobic power, in the safest possible way. Studies cited in that chapter will demonstrate that working out too intensely or overtraining may hurt your health by generating excess free radicals.

On the other hand, by gradually increasing your endurance through new, specially designed programs for fast walking, jogging, swimming, and cycling, you will minimize the danger that your body will produce excess free radicals during exercise. Also, your natural (endogenous) antioxidants—enzymes such as SOD and catalase—will become more powerful. As a result, those natural defenses will help you fend off the damaging work of any free radicals you encounter.

In chapter 5, you will embark on a special lower-intensity strength training program, which was designed to minimize your output of potentially injurious free radicals. After increasing your muscle strength, you will be less likely to become sore or injured when you face unexpected physical challenges. The soreness or fatigue that may follow an especially demanding game of tennis or lifting a heavy load is a sign that free radicals have been doing damage that may eventually result in the heart disease, cancer, or other disease described at the beginning of this chapter. That danger can be reduced or possibly eliminated by preparing your body ahead of time with the antioxidant strength program.

Step 2: Take a Daily, Specially Mixed Antioxidant Cocktail

Chapter 6 provides details on how to fight free radicals through recommended daily amounts of the three major antioxidant vitamins—vitamin C, vitamin E, and beta carotene. Also, I will discuss the mineral selenium and other supplements.

As you will learn, the antioxidant cocktail has a number of variations, depending on your age, gender, and activity level. For example, you may not know it, but if you are a fifty-year-old woman who walks regularly at an intensity level of 2.5 miles in 37:30 minutes, four to five days a week, you should be taking these daily doses of antioxidants: 600 IU of vitamin E; 1,000 mg of vitamin C; and 50,000 IU of beta carotene.

In general, whether they exercise or not, women over fifty years of age should take larger doses of vitamin E and beta carotene than younger women. And adult males at every age level should take more vitamin C than adult females of the same age. Men and women of any age who exercise heavily should take more vitamin C and vitamin E than men and women who are more sedentary.

It may be that you are one of those people who should not take an antioxidant because of certain side effects you experience. People who bleed easily, for instance, should avoid vitamin E. We will address that and other issues in chapter 6, as you design your own antioxidant action plan.

Step 3: Cook and Eat the Antioxidant Way

Chapter 7 presents a cooking and eating program which is based on principles developed at the Cooper Clinic. You will be given guidelines for using microwave and light steaming techniques to preserve the natural antioxidants in your vegetables, fruits, and other foods. Also, you will learn how to store and prepare foods effectively.

For example, the carotene content of green vegetables may be reduced by 15 to 20 percent through cooking, and the carotene value of yellow vegetables, such as corn, by 30 to 35 percent. Also, cutting up fruits into small pieces and soaking them after they are peeled can strip them of much of their vitamin C.

The more successful you are at preserving the natural antioxidants in your diet, the more powerful your internal antioxidant defense will become.

Step 4: Design a Life Relatively Free of Free Radicals

Chapter 8 provides you with an action plan to limit your exposure to environmental factors, including radiation, electromagnetic fields, and pollution, which may expose you to free radical damage. You will be encouraged to reevaluate your lifestyle so that you can embark on a life as free as possible of harmful free radicals.

As you move through those four steps in the following chapters, you may find at different times that you want more information on the scientific basis for using these practical techniques to prevent different diseases. Or you may want to refresh your memory about different terms or biological processes relating to free radicals and antioxidants. So I have included several extensive scientific appendices:

1. The Language of the Antioxidant Revolution—which focuses on the terminology and definitions that are important to an understanding of the work of free radicals and antioxidants.
2. The Scientific Foundations for the Antioxidant Revolution—the "internal plot" against your heart and blood vessels—a detailed description of how free radicals may trigger heart disease, cancer, cataracts, and other diseases.
3. The Benefits of Exercising in the Higher Ranges of a Fitness Program.
4. The Antioxidants in Your Food—how to preserve nutrients in the food you buy and prepare.
5. A handy reference chart, "Summary of Antioxidant Sources, Recommendations, and Effects."
6. A news release from the Council for Responsible Nutrition on "The Finnish Study: Puzzling Results."

Now you are ready to move on to the first step in the program—developing your own lower-intensity exercise regimen.

PART TWO

The Power of Lower-intensity
Exercise

4

The Lower-intensity Exercise Program

Recently, I was looking out of the window of my office, watching people working out at the Cooper Fitness Center on our one-mile track which winds around through trees, two ponds, and outdoor equipment. Most of those in the scene were my patients—and two of them in particular caught my eye.

The first was Tom, a top-level runner who ran frequent distance races, including marathons and masters competitions. Though he was forty-five years old, he regularly put in eight to ten miles of roadwork per day, six to seven days a week. In fact, he once told me that when he missed even one day of exercise, he felt guilty and tried to make up the distance he had lost by running farther and harder during his next couple of workouts. In addition to his endurance work, Tom spent at least a half hour, three days a week, doing strength exercises on the Cybex exercise equipment in our fitness center.

Tom's near-compulsive exercise habits had paid off in several ways. First of all, he was outwardly a near-perfect physical specimen, with well-toned and highly developed muscles and only about 6 percent body fat. Also, he experienced the satisfaction and self-esteem that accompany excelling at a demanding sport when many of one's peers are becoming couch potatoes.

My eye followed Tom's progress until he passed Bob, another of my patients, who was chugging along at a 9:30 mile pace. As Tom breezed by, Bob, a forty-three-year-old executive and family man, almost seemed to be standing still. But as I watched Bob I remembered how much my thinking on exercise had changed during the past five to six years. Bob jogs two miles a day, four days per week, and he does not come close to Tom in treadmill times or other fitness tests. Furthermore, his body fat hovers around 19 percent, and he definitely doesn't look like a highly trained athlete or body builder.

Which man comes closer to the ideal of exercising for the purpose of good health and fitness? Twenty years ago—perhaps even ten years—the answer would have been easy: Tom, the finely honed, middle-aged athlete would have received my vote without question. Today, I would choose Bob, primarily because of what I now know about how free radicals may damage the body through overtraining.

As a matter of fact, I was aware that Tom was probably suffering from free radical damage even as I watched him work out on that particular day. He was ailing from an injured and sore knee, a direct result of his putting in excessive running mileage. He was nursing sore upper body muscles and a tender shoulder joint from a heavy weight workout. His incidence of colds and viruses had been particularly high during the past year. And he had complained of excessive fatigue during his last medical exam. All of those symptoms and injuries have been associated in scientific studies with "distress" types of exercise and possible free radical damage.

Bob, in contrast, had no aches or pains; he had not experienced a cold or virus in more than two years, and his energy levels were as high as they had ever been. Something about his lifestyle was working in favor of better health. The secret, I had come to believe, was that his lower-intensity exercise program was bolstering—or at least not interfering with—his body's defense system.

Exercise Must Be a Centerpiece of Your Antioxidant Program

When you mention antioxidants, the first thing that usually comes to mind is vitamin supplements. But supplements are only one part of the antioxidant story—and only one step in the program that I am proposing in this book. What is often overlooked is that exercise must be at the center of any effective antioxidant action plan. The reason: Without regular exercise, your body's internal defenses against free radicals—including natural endogenous antioxidants like SOD, GSH, and catalase—may become too fragile for supplements to have their full effect.

But it is also essential to plug the right kind of exercise into your program. That means relying on *lower-intensity* exercise—which will minimize the output of extra free radicals as you work out and at the same time does the most to shore up your natural enzymes, or endogenous antioxidants. A number of recent studies have made the power of lower-intensity exercise clear.

What's the Evidence for Lower-intensity Exercise?

A July 29, 1993 report from the American College of Sports Medicine and the Center for Disease Control and Prevention came to this conclusion: "Every American adult should accumulate thirty minutes or more of moderate intensity physical activity over the course of most days of the week."

In making that recommendation, Dr. Walter Dowdle, acting director of the CDC, noted that a series of forty-three studies demonstrate that those who are inactive, or sedentary, are at almost twice the risk for cancer and heart disease as physically active individuals.

But think about it: an accumulation of just thirty minutes of moderate exercise per day. That is something any one of us can easily do—and just how easily, you will find out in my program.

Furthermore, the above recommendation is consistent with the guidelines given by Dr. Steven Blair, Director of Epidemiology at the Cooper Institute for Aerobics Research in Dallas. In his study, published on November 3, 1989, in the *Journal of the American*

Medical Association, Blair stated that thirty minutes of sustained activity, three to four times each week, would have a significant impact on reducing mortality from all causes. The results indicated that walking two miles in twenty-eight to thirty minutes, three times per week, is almost as beneficial as running two or more miles several times a week—if your main goals are to reduce risk of death from all causes and to prolong life.

To understand how Blair came to his conclusions, let's take a closer look at the study. Approximately 13,400 men and women were closely monitored for four years before any testing to ascertain that they were healthy. Then, they were asked to exercise to exhaustion on a treadmill. After age and sex adjustments were made, they were divided into fitness "quintiles"—or five groups— according to their treadmill times. Twenty percent of the participants were placed in each of the quintiles, and they were monitored for an average of 8.2 years after the treadmill tests to determine the cause of death of any participants.

At the conclusion of the study, the totally sedentary men, who were in the bottom 20 percent fitness category, had a death rate from heart disease, cancer, diabetes, and stroke that was 65 percent higher than the most active men in the top 20 percent category. But here is the interesting finding: Most of the benefits Blair discovered occurred when the men moved from the lower, totally sedentary 20 percent category to the next, moderately active 20 percent. Specifically, those in the lower 20 percent had a death rate that was 55 percent higher than those in the second, or next to lower, 20 percent. Women's results were comparable although not quite as dramatic.

In other words, there was not a straight-line decrease in deaths from all causes as men and women moved up the fitness scale. Instead the greatest drop-off in deaths occurred between the lower and next to lower categories of fitness, and then there was a very gradual drop-off in deaths as fitness levels increased further.

What is the significance of that study? First of all, a 55 to 65 percent reduction in deaths from all causes could translate to a two- to three-year increase in longevity. Another message—which has influenced the American Heart Association to make inactivity a primary risk factor for heart disease—is that the greatest benefits to health are experienced by those who move from complete inactivity just one step up, to a moderate, lower-intensity level of exercise.

But how exactly do you define "lower-intensity exercise" of the type that produces such significant health benefits?

What Do We Mean by "Lower-intensity Exercise"?

For purposes of your antioxidant action plan, I define "lower-intensity exercise" this way: The most effective exercise program for good health—including building up defenses against free radicals—is to exercise several times a week at your "target heart rate," which is a scientifically established higher-than-normal, but less-than-maximum rate that allows you to improve your endurance. You should maintain that level of exercise for at least thirty continuous minutes three times a week, or for twenty continuous minutes four times a week.

To determine your personal target heart rate, subtract your age from 220 to get your "predicted maximal heart rate." Then, take 65 to 80 percent of that figure to obtain your target heart rate.

For example, a forty-year-old would have a predicted maximal heart rate of 220 minus 40, or 180. Multiplying that figure by 0.65 and by 0.80 produces a target heart rate range for endurance exercise of 117 to 144 heartbeats per minute.

To get a better idea of how a lower-intensity program may work in practice, consider the following illustrations, which are based on an actual program used by many of my patients.

A Model Lower-intensity Exercise Program

A simple method to set up a lower-intensity exercise program is to use my fitness points method—which is an easy way to measure the degree, intensity, and amount of exercise you achieve during a given workout. If you work up to a total of at least fifteen points per week, using one or a variety of endurance exercises, you will ensure considerable benefit from lower-intensity exercise.

For example, if you walk two miles in less than thirty minutes three times per week, you will earn those fifteen fitness points. Or if you walk three miles in less than forty-five minutes two times per week, you will earn sixteen points. Either approach will give you the amount of physical activity you need to maximize your health and longevity and minimize your production of destructive free radicals. Here is another possibility: You might walk two miles in less than forty minutes five times per week for fifteen points. As you can see, there are many acceptable variations. (See the descriptions of specific exercises at the end of this chapter for more possibilities.)

How about Gardening?

Some have argued that it is acceptable to incorporate fragmented or interrupted activities, which may be done a few minutes at one time of the day and a few minutes at another time. The ultimate goal, according to this view, is just to be sure that you spend at least thirty minutes a day in some sort of physical activity.

For example, a person takes a few minutes to walk up a flight of stairs instead of taking an elevator. Then she spends a few more minutes doing some gardening or raking leaves at home after work. At night she goes out dancing. As this argument goes, the time devoted to each activity may be added to make up the thirty-minute total each day. Furthermore, the proponents of this position would say that there is nothing wrong with cutting back on your current exercise program, even if you are involved with a relatively moderate regimen.

My main reservation with this point of view is that it is extremely difficult to quantify the effect of sporadic activities in the same way we can quantify continuous exercise like walking or cycling. Without that ability, it is hard to know exactly how much benefit you are getting or to check how much, if at all, your fitness level is improving or deteriorating. But even interrupted or fragmented physical activity is better than nothing.

The findings about the tremendous benefits of lower-intensity exercise suggest that we should begin to look at physical fitness—and the ways we classify our conditioning activities—in a totally different way.

A New Way of Looking at Fitness

The Cooper Institute for Aerobics Research study mentioned above, as well as the studies we will discuss shortly that link overtraining to free radical production, has convinced me that we should stop pegging people as simply "fit" or "unfit." Rather, we should use three separate classifications for inactivity and exercise, which in ascending order of physical conditioning can be stated like this:

1. The sedentary level of little or no physical activity.
2. The health and longevity fitness level of lower-intensity exercise.
3. An athletic level of exercise—which approaches and some-times even reaches the level of fitness required for competi-tion, but which *does not* involve overtraining.

Those at the first level will be likely to have more disease, live shorter lives, and experience a relatively low quality of life. Those at the second level will enjoy significantly higher benefits than those at level one, in terms of lower disease rates and longer lives.

Most of those at the third or highest level of fitness will be characterized by slightly longer lives than those at level two; slightly lower risk of serious illness than those at level two; and considerably higher quality of life than those at level two. (For more information on the many benefits of exercising at the upper levels of fitness, see appendix 3.)

Note: Remember that even the third and highest category refers to exercise programs that are designed to minimize the output of free radicals. This level should not be understood to encompass excessively rigorous regimens that often are accompanied by over-training. Those who push themselves beyond what is necessary to enjoy the full benefits of aerobic fitness—including many mara-thoners and other highly competitive athletes—may lose the very benefits for which they are striving. As you will see in a moment, such loss of health through overtraining can be traced directly to the free radical phenomena (which may be controlled by increas-ing antioxidant defense).

An excellent way of pursuing intelligent exercise—with a goal of either health and longevity fitness or athletic fitness—is through a well-designed walking program. Recent studies at the Cooper Institute for Aerobics Research have provided the basis for such a program.

Why Fast Walking May Be Your Best Bet

One of the new features of the exercise options in this book is the description at the end of this chapter of the fast walking pro-gram (also called "endurance walking," or at the competitive level, "race walking"). With this approach, you can achieve optimum endurance benefits *without* stimulating excess free radicals and *without* risking the incapacitating muscular and skeletal problems that are so common with higher-intensity exercises. In other words, you are unlikely to suffer such injuries as strained muscles, pulled ligaments, or joint and knee problems.

These new discoveries about the benefits of fast walking came from a study performed by Dr. John Duncan, an exercise physiologist at the Cooper Institute for Aerobics Research. His investigation, the results of which were published in the *Journal of the American Medical Association* on December 18, 1991, involved the monitoring of 102 premenopausal women for a total of six months. During that time they were divided into one control group, who did not change sedentary daily habits, and three walking groups.

The walkers were encouraged to walk three miles, five times per week, but each of the three exercising groups was assigned a different rate of speed in which to cover that distance. One group was told to do it at a pace of twenty minutes per mile; the second group was to go faster, at fifteen minutes per mile; and the third group was to move at the fastest clip, twelve minutes per mile. Because the rate of the speed of the last group was so demanding, they were given seven weeks of training under the guidance of an expert walker, Casey Meyers, author of *Aerobic Walking*. That way, they were able to achieve the goal of walking three miles in thirty-six minutes on a regular basis.

Duncan's study soon revealed that a twenty-minute-per-mile pace produced an average exercise heart rate that was 55 percent of the maximum predicted rate. The fifteen-minute-per-mile rate caused the heart rate to go up to 68 percent of the predicted maximum. And the hearts of the twelve-minute-per-mile walkers hit 86 percent of their predicted maximum.

After six months of such exercise, the fitness of the walkers, as measured by their ability to process oxygen during exercise, increased in these amounts:

- In the twenty-minute milers, there was a 4 percent increase in endurance fitness.
- In the fifteen-minute milers, there was a 9 percent increase.
- In the twelve-minute milers, there was a 16 percent increase. Furthermore, those walking the fastest, at the twelve-minute pace, achieved all the fitness benefits that they would have gained if they had been jogging nine-minute miles.

Perhaps the most surprising observation was that even though the women collectively covered more than twenty thousand miles during the six months of the study, there was *not a single musculoskeletal problem* that required medical attention. If that had been a running/jogging study involving people in the same age group, I can assure you that many of the participants would have developed severe, if not incapacitating, musculoskeletal problems. In other words, fast walking—a classic lower-intensity exercise—can be just as effective as jogging, and you will have fewer injuries of the muscles, bones, and joints.

Caution: If you walk faster than twelve minutes per mile, that becomes true race walking. Even though this activity will produce higher levels of fitness, it may also be associated with an increase in musculoskeletal problems. For that reason, I recommend that race walking at the sub-twelve-minutes-per-mile rate be limited to the competitive athlete. (More details on how to incorporate a lower-intensity, fast-walking program into your regimen are available later in this chapter.)

Underlying the problems with competitive race walking, as well as other high-intensity exercise, is the specter that was introduced at the beginning of this chapter—the threat of free radicals. Or to put it another way, very rigorous conditioning programs including different forms of overtraining may actually backfire on your body.

The Free Radical Phenomenon

Long before we reached our present understanding about how free radicals can harm human health, I cautioned that if the goal was simply cardiovascular fitness, it was unnecessary for anyone to jog more than twelve to fifteen miles per week. Obviously, if you hope to engage in distance races or if you want to develop the levels of endurance necessary for other competitive sports, you need to train more than that minimum. But for the average person, shorter distances are quite adequate for full health and fitness benefits.

Now there may be even more pressing reasons to limit the intensity of your exercise—reasons that center on the danger posed by excess free radicals, which may be produced during heavy training programs. Exhausting, high-intensity exercise may increase your susceptibility to different cancers, heart attacks, cataracts, premature aging, decreased immunity, and a variety of other medical problems.

What is the evidence for this charge that I am leveling against high-intensity exercise programs? Let's look back a few years to a study done by Dr. Ralph Paffenbarger and his associates of 16,936 male Harvard alumni, aged thirty-five to seventy-four years. According to a report in 1986 in *The New England Journal of Medicine,* he found that death rates were lower for men who were involved in regular physical activity than for men who were not. The death rates declined steadily as the number of calories they burned per week increased. In fact, the rate of death was one-fourth to one-third lower among men expending two thousand or more calories per week, as compared with the less active men—a finding that translates to an estimated increase in life span of 2.4 years.

But at the highest intensity levels, exercise seemed to be less beneficial. The rates of death began to go up slightly among men expending more than three thousand calories per week. (Epidemiologists refer to this finding as the "reverse J-slope phenomenon" because of the way the results look on a graph. The rate of death slopes downward at first as exercise intensity goes up. But when the intensity of physical conditioning reaches its highest pitch, the death rate goes up slightly. The mortality graph reflects this fact as the downward sloping line suddenly turns back up, to form the last stroke of a reverse "J.")

Paffenbarger's findings were reflected in Dr. Steven Blair's study from the Cooper Institute for Aerobics Research, which was published in 1989 in the *Journal of the American Medical Association.* In that investigation, there was a slight increase in the death rate from all causes among women who were engaging in the heaviest workouts.

Those findings may possibly be explained by reports in the scientific literature, which have established that free radical production increases during exercise—and that such free radical activity is associated with oxidative damage in the muscles, liver, blood, and other tissues.

Of course it is often difficult to find conclusive, undeniable data when dealing with human subjects. Even though the experts' consensus is that high-level, "ultra"-type exercise over long periods may actually increase the risk of free radical damage, some studies indicate that athletes who have trained in that fashion may actually live longer.

A 1993 article by Seppo Sarna and fellow researchers in *Medicine and Science in Sports and Medicine* reports a study of 2,613 world-class Finnish athletes. They had participated in the Olympic Games, world championships, or European championships during the years 1920 to 1965. They were compared with 1,712 conscripts inducted into the Finnish Defense Forces during the same time period.

The researchers found that the greatest life expectancy, adjusted for occupation, marital status, and age, was in the endurance athletes. The increased life expectancy was explained mainly by decreased cardiovascular mortality.

Another interesting finding, however, was that the former athletes who were still alive in 1985 had nearly all continued to be physically more active than the conscripts to whom they were compared. Furthermore, those athletes who continued to be physically active after their competitive years enjoyed a greater life expectancy than those athletes who had been highly trained only during their early years. That finding gives credibility to the theory that lower-intensity, regular, long-term activity may enhance the antioxidant defense system, whereas a high-intensity, short period of activity may actually decrease the body's defenses.

I constantly encounter situations with patients that confirm those findings. In particular, I have become alarmed at the increasing frequency of atrial fibrillation (irregularity of the heartbeat) in highly conditioned runners who have been exercising over a period of many years. Also, I am bothered by the frequency of cancer of the prostate among my older patients who are marathoners and "ultra" athletes.

At this point, we have no studies that allow us to make definitive comparisons of these patients with control subjects who do not exercise as intensely. But my clinical observations make me suspicious of overtraining and—with the mounting evidence in the scientific literature—cause me to recommend a more moderate approach for many amateur athletes.

The Link Between Overtraining and Free Radical Damage

During normal conditions free radicals are generated at a low rate and neutralized by the body's well-developed scavenger and antioxidant systems. But if a greatly increased rate of free radical production is triggered by overtraining, the number of radicals may exceed the capacity of your cellular defense systems. The unrestrained extra outlaw molecules then launch attacks on your cell membranes, with a resulting loss of cell viability and an increase in skeletal and muscle damage. The damage to and inflammation of tissues that often accompany exhaustive exercise are the most obvious signs of free radical activity.

In people who are out of shape or untrained, strenuous exercise also causes the appearance of "oxidant stress markers" in the blood and muscle tissue. Special laboratory tests are used to uncover these markers. On the other hand, studies have shown that trained muscles are resistant to oxygen stress unless the exercise is so heavy and prolonged that it challenges the muscle's glycogen (sugar) supply.

How Does Your Body Produce Extra Free Radicals During Exercise?

There are at least two ways that free radicals are produced during exercise. The first involves an electron leak, which may occur during exhaustive exercise. The increase in the body's oxygen consumption during a heavy workout can increase ten to twentyfold or more. In fact, in individual muscle fibers that are put under the greatest stress, the oxygen use may go up by one to two hundred times above normal. That tremendous pumping of oxygen through the tissues triggers the release of free radicals in those tissues. Also, during such exercise the output of free radicals—specifically the superoxide radical—may soar in the mitochondria, or the energy centers of cells.

The second way that free radicals are produced during exercise relates to a process known in medical terms as "ischemia reperfusion." Here is how it works.

When you exercise intensely, the blood flow in your body is shunted away from the organs that are not actively involved in the exercise process, such as the liver, kidneys, stomach, and intestines. Instead, the blood is diverted to the working muscles, including the heart and legs. During the shifting of the blood flow, a part or all of the body regions or organs not involved in exercise will experience an acute lack of oxygen (known as "hypoxia").

When the exercise is completed, blood rushes back to the organs that were deprived of blood flow. That process, known as "reperfusion," has been linked to the release of an excessive number of free radicals.

There is another side to this reperfusion phenomenon. Even muscles that are actually involved in heavy exercise may undergo some degree of oxygen deprivation, especially when the workout approaches or reaches the level of exhaustion. Known as "maximal oxygen intake," or "exhaustive exercise," this type of activity can occur if you sprint or otherwise push your body to its limits. You reach a point at which you cannot process enough oxygen to keep going, and oxygen starvation of your muscles and organs begins to occur.

Again, when you stop this type of exercise, blood rushes back into the organs that have been deprived of oxygen, and free radicals explode into your tissues. A number of studies have shown such excessive free radical production in skeletal muscles and livers of animals during exhaustive exercise.

Fortunately we have an internal antioxidant defense system that helps protect us from free radicals and the "oxidative stress" that they produce. You will recall that your body produces enzymes such as SOD (superoxide dismutase), which was discovered in 1968 by American scientists J. M. McCord and I. Fridovich (see chapter 2). Scientists have also identified other internal, or "endogenous," antioxidants, such as catalase and glutathione peroxidase (GSH).

But high-intensity training may overwhelm these internal defense mechanisms. Your body's natural "police force" may break down. In such situations, oxidative damage will occur and may affect every part of exposed cells, including the cell protein, the lipids (such as cholesterol), and the cell nucleus. That is the sort of damage that eventually may lead to cancers, coronary artery blockage, and other diseases.

These dangers have prompted scientists to accelerate their research into the measurement of free radical activity in different types of exercise.

Measuring Free Radicals in Exercise

One way to evaluate whether you are exercising too intensely is to find out whether or not your level of activity is producing excess free radicals. Unfortunately, it is not easy to measure free radicals directly. They move too quickly and last only a fraction of a second in an independent state before they lock on to another molecule and begin the oxidation damage. So scientists have resorted to the next best thing: they measure the "footprints" or residues left by the radicals.

One method of taking these measurements is to check the amount of pentane, a residue of free radicals found in expired air from breathing. In an early experiment in 1928, a group of people exercised on a bicycle ergometer for twenty minutes at 25 to 50 percent and at 75 percent of their maximal performance ability. At the lower intensities of exercise, there was no increase in the expired pentane from their breathing. But twenty minutes of exercise at 75 percent of their maximal performance caused a nearly twofold increase in the expired pentane—a sure indication that their bodies were putting out excess free radicals.

A second method of measuring free radicals focuses on the process of oxidation of cell fats, known as "lipid peroxidation." This process, which is caused by free radical damage, produces residues known as thiobarbituric acid reactive substances, or "TBARS." Through special lab blood-testing procedures researchers can measure TBARS and estimate the free radical damage that is occurring.

Such techniques have helped scientists ascertain that heavy exercise does indeed lead to free radical damage in nonhuman subjects. For example, researchers K. J. A. Davies, Lester Packer, and their associates reported in 1982—and H. M. Allesio and associates confirmed in the *American Journal of Physiology* in 1988—that after intense exercise, animals show an increase in TBARS in their muscles and livers.

The findings with human beings present a more complex picture. One 1988 study on ultramarathoners, who raced for an 80.5 kilometer (fifty-mile) distance, revealed increases in their levels of TBARS. Also, an investigation by researcher L. Viinikka and his colleagues in the 1984 *Medical Science Sports Exercise Journal* showed an increase in TBARS with exercise, even though the athletes participating were highly trained.

On the other hand, a 1990 report in the *Archives of Biochemistry and Biophysics* found that runners who completed a half-marathon (21 kilometers, or 13.1 miles) experienced no change in the TBARS in their blood.

Why was there no apparent residue from free radicals with these athletes? Most likely, the highly trained condition of these competitors and the shorter distances they were running combined to keep their free radical production at low levels.

Similar findings emerged in a 1987 report by the scientist R. Lovin and his colleagues in the *European Journal of Applied Physiology*. These researchers discovered a decrease in lipid peroxidation (free radical cell damage) in people who exercise at 40–70 percent of their maximal capacity, whereas there was an increase in this damage among those who worked out at 100 percent of their capacity.

Eric Witt, Lester Packer, and other experts at the Department of Molecular and Cell Biology, University of California at Berkeley, summed up the situation quite well in 1992 in the *Journal of Nutrition:*

> The chances of finding evidence for oxidative damage during exercise seem to depend on the intensity of the exercise, the location of the sample site and the state of training of the subjects. Intense or exhaustive exercise in untrained subjects is more likely to produce oxidative damage, which is more likely to be seen in muscle than in blood.

Their conclusion finds support in a study we have been conducting at the Cooper Institute for Aerobics Research. In 1993, Dr. Neil Gordon, the coordinator of the study, selected ten highly trained men and women and ten sedentary men and women who had not consumed any antioxidant supplements in the six weeks preceding the test. Our objective was to determine if the level of exercise training would have an effect on the body's free radicals in a situation where the body was not protected by antioxidants.

Half of the men were highly trained, well-conditioned athletes who ran an average of twenty-two miles per week. The other half of the men were totally sedentary. Half of the women were well-trained—but not as intensively trained as the trained men. Specifically, the trained women were running an average of ten miles per week on a regular basis at the time they were tested. The other half of the women were totally sedentary.

As our measure of free radical activity, we chose to evaluate the TBARS present in the blood of the participants. On three separate occasions, approximately one week apart, all of the subjects were evaluated in a completely resting state.

On two additional occasions, the groups were exercised to exhaustion on a motor-drive treadmill. Blood specimens from the participants were collected immediately after the conclusion of the exercise, one hour after the exercise, six hours after the exercise, and twelve hours after the exercise.

In each evaluation, the highly trained male athletes had the highest level of TBARS, a result that indicates a relatively high level of free radical activity. The moderately trained women, in contrast, had the lowest levels of TBARS, or free radical activity. The untrained men and women fell in the middle, with TBARS results between those of the trained men and the trained women. There was no significant change in the TBARS findings at any of the evaluations of the four groups, whether they were tested at rest or after exercise.

Here are the specific results of our study:

Group	Exercise	TBARS
Women	Moderately trained	1.57
Men	Untrained	1.71
Women	Untrained	1.82
Men	Highly trained	2.32

What are we to make of those preliminary findings? Most likely, the explanation lies in the benefits of moderate, lower-intensity exercise.

The group with the lowest TBARS results, and hence the lowest levels of destructive free radical activity, was the moderately trained women. By pursuing regular, lower-intensity exercise, they apparently avoided stimulating excessive free radical production in their bodies and at the same time bolstered their internal antioxidant enzyme defenses. (As you know, strong endogenous antioxidant activity will help quench harmful free radical attacks.)

The group with the highest TBARS results, and thus the highest levels of destructive free radical activity, was the highly trained men. By doing heavy, "distress" type of exercise, they tended to stimulate excessive numbers of free radicals, which their internal antioxidants were unable to handle.

The middle groups—the untrained men and women—seemed to avoid the extra free radical attacks confronted by the highly trained men. But at the same time, they do not enjoy the protection afforded the women who were on a lower-intensity exercise program.

The main principles gleaned from these findings are fairly simple and straightforward. To exercise at the safest levels and protect yourself from free radical damage as much as possible, you should:

- Utilize the antioxidant cocktail as recommended in chapter 6 to help neutralize any potentially harmful effect of excess free radicals.
- Avoid exercises or activities that place an undue or prolonged strain on your body or its oxygen production systems, such as the heart and lungs. In other words, do not push yourself to do frequent, high-intensity exercise, particularly to the point of total exhaustion and chronic fatigue.
- Train regularly on a lower-intensity exercise regimen. Such a program will maximize your inner antioxidant defenses and avoid the production of free radicals that can be so damaging to your health. By gradually becoming more fit, you will be better able to handle those unexpected stresses and strains that you are likely to encounter in everyday life.

But in practical terms, what is the upper end of safe exercise? Or to put it another way, what should your most effective lower-intensity exercise program look like?

Designing Your Personal Program

In setting up your personal lower-intensity endurance exercise program, keep in mind three basic guidelines. (Chapter 5 contains guidelines for those who also plan to incorporate a strength program into their workouts.)

1. Exercise at a heart rate that will minimize your output of excess free radicals.

From my clinical experience and my study of scientific findings on this subject, I recommend that you not allow your heart rate to exceed 80 percent of your predicted maximum heart rate. This is the ceiling above which free radicals tend to be produced at excessive levels.

To calculate your "ceiling" heart rate figure, first determine your predicted maximum heart rate by subtracting your age from 220. Then, multiply that result by 80 percent (0.80) to get the top heart rate at which you should exercise.

For example, if you are a forty-year-old man, your predicted maximum heart rate would be 180 heartbeats per minute (220 minus 40). Then, you would multiply 180 by 0.80 to get 144 beats per minute—the maximum intensity at which you should exercise in order to control your output of free radicals. By applying this first guideline you can monitor the intensity of your effort, no matter what kind of exercise you do. Simply stop briefly during your workout and place a finger (not the thumb) against one of your carotid arteries, which are on the front sides of your neck, just to the right and left of your Adam's apple. (Caution: Place a finger against only *one* of the arteries; pressing on both of them could stop the flow of blood to your brain and cause you temporarily to lose consciousness.)

When you feel your pulse, look at the second hand on your watch and count off the number of times your heart beats in ten seconds. (Begin counting with "zero" and end with "ten.") Multiply that result by six to get the number of heartbeats per minute.

By taking your pulse in the middle of a workout, you will be able to ascertain fairly accurately just how hard you are working out—and whether you need to lower the intensity of your effort.

2. You can also keep yourself safe from excess free radicals by earning no more than fifty fitness points per week. Translated into actual activities, a fifty-plus point per week workout might look like this:

- A jogger under thirty years of age, who jogs three miles in twenty-four minutes (an eight-minute-per-mile pace), five times a week, would earn eighty-five points.
- A forty-year-old exerciser jogging on a treadmill at a speed of 6 m.p.h. with a flat incline for thirty minutes, five days a week, would earn seventy points.
- A fifty-five-year-old jogger who jogs three miles in thirty-two minutes, five times a week, earns fifty-five points.

Each of those regimens is above the fifty-point threshold that represents the level at which free radical activity tends to become excessive. You should either exercise at a lower-intensity range so that you earn fewer than fifty points, or if you choose to work out more rigorously, be sure to take the extra antioxidant supplements as prescribed in chapter 6 for heavy exercisers.

What is the scientific authority for my establishing this threshold? Dr. R. S. Paffenbarger's study of Harvard graduates has shown that those who exercise at no more than a fifty-point-per-week level have excellent protection against disease. The 1989 study conducted at the Cooper Institute for Aerobics Research by Dr. Blair has confirmed that conclusion.

Still another study reported in the January 1993 issue of *Age and Aging* that nine of fifteen men participating in an exercise study had "some form of pathological disease" by the average age of seventy. The nine injured or ill men, who were observed for twenty-five years, had to stop exercising because of a variety of complaints, including damaged knees, lymphoma (a type of malignant or pre-malignant tumor), and atherosclerosis, which in one instance required bypass surgery.

How hard had the sick men been exercising? They had been accumulating a total of seventy-two to ninety-seven fitness points per week, well in excess of the fifty-point recommended maximum.

3. Follow the directions in the charts at the end of this chapter.

If you choose to use one of the traditional sports for your exercise, such as walking, jogging, cycling, or swimming, all you have to do is turn to the charts at the end of this chapter to find a progressive program to suit you.

But before we move on to the actual exercise program, let us consider our last topic—the use of antioxidant supplements for those who are involved at all levels of intensity of exercise.

Combining Antioxidant Supplements with Exercise

You will note in the following exercise charts that I have included suggestions for relatively high amounts of antioxidant supplements for those who work out systematically. The use of relatively high amounts of antioxidant supplements is especially important to afford protection to those who exercise near or above the exercise intensity "ceilings" described in the previous section.

If you are a competitive athlete of any age, for example, you will undoubtedly find yourself earning more than fifty fitness points per week. Also, it is likely that your heartbeat during exercise will rise above 80 percent of your predicted maximum heart rate. As we have seen, exercising at levels above those thresholds may greatly increase the risk of your body tissues suffering from excess free radicals. In such situations, antioxidant supplements can provide important protection against danger.

The body's increased need for oxygen during exercise seems to produce free radicals, which can oxidize the fats in muscle cell membranes. That process, called "lipid peroxidation," may make cells more susceptible to aging and other damage.

66

Since antioxidants can disarm free radicals, exercisers should be able to protect themselves by taking the right kind of supplements. Research conducted at the University of Washington School of Medicine in St. Louis has provided credibility to such a theory. Every day for six months, eleven young men were given 600 IU of vitamin E, 1000 mg of vitamin C, and 30 mg (50,000 IU) of beta carotene. Nine other men were given a look-alike but inactive supplement (i.e., a placebo).

At the start of the study, all the men ran on a treadmill for 35 minutes and then had the free radicals they produced measured. When they were tested six months later, the subjects who took antioxidants formed 17 to 36 percent fewer free radicals than those taking placebos. (See *Nutrition Action Health Letter*, Sept. 1993, and the *Journal of Applied Physiology*, 1993, Vol. 74, p. 965.)

This study provides strong evidence for using antioxidant supplements with exercise. The antioxidant supplements most often used by exercisers include vitamin E, vitamin C, beta carotene, the mineral selenium, and the coenzyme Q10. Scientists and physicians in clinical practice have observed benefits from antioxidant supplementation among patients with a number of pathological conditions associated with excess free radicals, such as hypoxia (oxygen deprivation), ischemia (deprivation of blood supply to tissues), and reperfusion injuries (the sudden return of blood to tissues such as the heart which have been temporarily deprived of blood).

Most of the studies that have examined free radical damage have focused on vitamin E, vitamin C, and selenium. For example, it has been established that the naturally produced antioxidant glutathione peroxidase (GSH) is weakened or reduced in the body by a deficiency of the mineral selenium. I regard taking selenium supplements in the amounts of 50 to 100 micrograms daily as an acceptable but optional practice. The reason for my conservatism is that at this point, scientific studies have not firmly established that supplementation helps increase the body's selenium defenses to a significant extent. Also, selenium overdose has been associated with hair loss and other toxic effects. But cautiously I continue to recommend small amounts of selenium as an optional supplement because of the possible association of the mineral with benefits to GSH.

Why is GSH so important? As a part of your body's internal (endogenous) antioxidant "police force," GSH counteracts the harmful effects of hydrogen peroxide. Also, GSH helps control the process of lipid peroxidation, which may cause the "bad" LDL cholesterol particles to oxidize. You will recall from our prior discussion that oxidized LDL contributes to the formation of "foam cells," which, in turn, become the foundation of blood vessel plaque, clogging of the arteries, and heart attacks.

Scientific evidence is more solid for the benefits of taking vitamins C and E in dietary or supplement form. First of all, keep in mind that vitamin E is found in LDL cholesterol and serves as the major defense system against free radicals that threaten to turn LDL particles into foam cells. And remember, vitamin C enhances the antioxidant effects of vitamin E.

My consultants Drs. Scott Grundy and Ishwarlal Jialal of the University of Texas Southwestern Medical Center at Dallas have established in a laboratory setting that vitamins C and E can prevent oxidation of LDL, which may help to prevent atherosclerosis. In addition, a variety of studies on animals have shown that those deficient in vitamin E may have a sixfold increase in pentane in their expired breath (a measure of free radical activity), as compared with animals that have sufficient stores of vitamin E.

In humans, most of the scientific studies support the idea that vitamin E supplementation protects the body against exercise-induced free radical damage. In a 1987 evaluation by researcher Pincemail and colleagues, expired pentane values were dramatically lowered by vitamin E supplementation.

A 1993 report in *Medicine and Science in Sports* showed that five minutes of 100 percent effort produced a marked increase in expired pentane production twenty minutes after the exercise was completed. But when the participants were given 200 milligrams of vitamin E daily for three weeks, the pentane production during this extreme high-intensity exercise was reduced by 75 percent.

In still another study, done by C. J. Dillard and fellow researchers in 1978 and reported in the *Journal of Applied Physiology*, taking 600 milligrams of synthetic vitamin E (dl-alpha tocopherol) three times daily for two weeks reduced the participants' pentane production during exercise. In that investigation, the exercise was performed at 75 percent of the participants' maximal capacity.

You should be aware that synthetic vitamin E is indicated by the "l" after the "d," whereas natural vitamin E is indicated by an absence of the "l." So what about the effect of natural vitamin E (d-alpha tocopherol)?

A 1989 study, conducted by S. Sumida and colleagues and reported in the *International Journal of Biochemistry*, directed exercising participants to take 300 milligrams of natural vitamin E daily for four weeks. (That amount is about the same as 300 IU of vitamin E.) The findings? When the exercisers took the natural vitamin E, they had a lower production of TBARS than when they did not take the vitamin.

Remember: As I cite these results, whenever there is an increase of pentane in the expired breath or an increase of TBARS in the blood, that is evidence that free radicals have been about their damaging work.

We can see that antioxidant supplements may have a direct effect on reducing free radical damage after exercise. But can they go a step further? Can they improve athletic performance as well?

Can Antioxidants Improve Athletic Performance?

At least one group of athletes—mountain climbers—have fared fairly well using vitamin E supplementation. In a 1988 study by researchers I. Simon-Schnass and H. Pabst, a group that was given

200 IU of the supplement twice a day did not exhibit the same deterioration in performance during a hard workout as a group that did not· receive the vitamin. According to the researchers' report in the *International Journal of Vitamin and Nutrition Research*, the supplemented group also had a smaller increase in pentane content in their expired breath.

In other cases, taking antioxidant supplements does not seem to have had much effect on the ability to turn in a better performance during exercise. For example, a group tested by researcher S. Sumida in 1989 was monitored as they exercised to exhaustion, both before and after taking vitamin E supplements. That group exhibited no difference in their ability to process oxygen or in their exercise times as a result of taking the supplements. Similarly, in studies involving swimmers, one group that took vitamin E supplements did not perform any better than another group that only took placebos.

At present, there is no evidence that other antioxidant supplements—such as vitamin C, beta carotene, selenium, and coenzyme Q10—improved athletic performance. To sum up, then, supplementation, particularly with vitamin E, may provide some boost to the performance of some athletes. But even if you are not able to increase your speed or stamina with supplements, additional supplementation is always a good idea to prevent oxidative damage. So specifically, what supplements should the very active individual take?

Antioxidant Recommendations for the Active Individual

As we have seen, there is strong support for the position that vitamin C enhances the effect of vitamin E. Also, beta carotene has been linked to a lower incidence of lung cancer and other free radical-related disease. Exercise of any type will increase your free radical production. So antioxidant supplements are *mandatory* for adults involved in both types of exercise: (1) the health-longevity fitness exercises, and (2) the athletic fitness exercises (involving the achievement of fifty or more fitness points weekly). Those working out in the lower, health-longevity range should follow supplement Recommendation #1. Those involved in the more demanding and rigorous ranges of the athletic fitness program should follow supplement Recommendation #2. The supplements should be taken in these amounts:

Recommendation #1: For Health and Longevity Fitness

Daily antioxidant doses for those who are exercising at a level of intensity that is *less* than 80 percent of their predicted maximum heart rate, or at a level below fifty fitness points per week:

- Natural vitamin E (d-alpha tocopherol): 400 IU for those twenty-two to fifty years old; 600 IU for those over fifty years of age
- Vitamin C (ascorbate): 1,000 mg (500 mg taken twice daily) for women; 1,500 mg (750 mg twice a day) for men twenty-two to fifty years old; 2,000 mg (1,000 mg twice a day) for men over fifty
- Beta carotene: 25,000 IU for those twenty-two to fifty years of age; 50,000 IU for those over fifty

Note: These recommendations are the same as those for non-exercisers (see chapter 6).

Recommendation #2: For Athletic Fitness

Daily antioxidant doses for those exercising at a level of intensity that is more than 80 percent of their predicted maximum heart rate, or at a level of fifty fitness points per week or higher:

- Natural vitamin E (d-alpha tocopherol): 1,200 IU
- Vitamin C (ascorbate): 2,000 mg for women; 3,000 mg for men (split evenly into two or three doses per day)
- Beta carotene: 50,000 IU

Note: These recommendations are the same as those for people who weigh more than two hundred pounds (see chapter 6).

As I have already indicated, the evidence for taking selenium as a regular supplement is scarce, so I regard that as optional. Taking daily doses of 50–100 micrograms is acceptable for those who want to add this supplement to their regimen and should not cause symptoms of overdosage.

The coenzyme Q10 is the "newest kid on the block" as a possible antioxidant supplement. There are only a few studies that have suggested the beneficial effects of this enzyme in preventing exercise-induced oxidative stress, and what evidence we have is far less weighty than that supporting vitamin E. Among researchers who are exploring this area, it is agreed that coenzyme Q10 might help enhance the power of vitamin E in protecting cell membranes from oxidative damage from free radicals, much as vitamin C does. At this point, however, because of the scant scientific support for its benefits, I do not recommend that my patients take coenzyme Q10.

Now, armed with this information supporting the importance of lower-intensity exercise—and also with specific recommendations for antioxidant supplements for exercises—let us move on to a description of the actual programs. In the remaining pages of this chapter, you will find charts and explanations for these programs and charts:

- The fast walking program
- The fast walking aerobics points charts
- The treadmill walking program
- Five sample health and longevity fitness programs—regular walking, fast walking, walking-running, cycling, and swimming
- Five sample athletic fitness programs—walking, fast walking, walking-running, cycling, and swimming.

Caution: You should always undergo a thorough medical examination by a qualified physician before you embark on a new exercise program such as this one. If you have a history of medical problems, particularly heart disease, medical clearance is an absolute requirement.

Now, choose the program that most appeals to you and start enjoying the higher energy and fitness levels of the Revolution! We will begin with an in-depth description of my all-new fast walking program, which has been designed especially for this book.

The Fast Walking Program

Fast walking—also known as "speed walking" or in a competitive sense, "race walking"—has been gaining in popularity because of its effectiveness in promoting endurance with minimal injuries. Comparative studies indicated these comparable energy expenditures for aerobic walking and running:

Walking	Running
11:00 min./mile =	8:00 min./mile
12:00 min./mile =	9:00 min./mile
13:00 min./mile =	10:00 min./mile
14:00 min./mile =	11:00 min./mile

Fast walkers, or "speed walkers," actually burn off more calories per mile than joggers because of their shorter strides, the necessity of taking more steps to cover the same distance, and the required vigorous movement of their arms. At a pace of twelve minutes per mile, fast walkers burn 530 calories per hour, whereas joggers at the same pace burn only 480 calories per hour. In other words, it is a more vigorous workout to walk at a 12:00-minute-per-mile pace than to jog at that speed. If you don't believe this, just try it!

A significant advantage of fast walking is that it causes less wear and tear on the body. In proper foot gear, walking is a shock-absorbent sport, whereas jogging tends to put stress on the heels, knees, and back, and increases the chances of injuries.

But you have to prepare to be able to walk at the faster speeds. As mentioned earlier in this chapter in the Duncan study on women walkers, it takes six to eight weeks to learn the walking techniques and to be able to walk for two to three miles at a 12:00-minute-per-mile pace.

As with any regular exercise it is important to stretch and warm up for three to five minutes. The best way to do this is to walk at a moderate pace during the warmup period and also to incorporate exercises to stretch the major muscle groups. That may include slow hamstring stretches, Achilles tendon stretches, and side bends. (For a full description of different stretching exercises, see the flexibility program in chapter 5.)

After you have warmed up, begin your program by walking at a 20:00-minute-per-mile pace—a rate that translates to about ninety steps per minute. As you walk, use the correct technique as described below.

- Keep your back straight, with head held high, and chin almost in line with the shoulders.
- Bend your arms at a right angle at the elbow and let them swing parallel to the body or at a slight diagonal across the body.
- The speed of your arm movements should be related directly to your leg movements and can enable you to increase your walking speed. Pump your arms to add momentum to your walk, but keep the arms relaxed and avoid exaggerated arm movements.
- Your lead heel should make contact with the ground before your rear toes leave it. In other words, one foot must always be in contact with the ground. Otherwise, you are jogging or running!
- Your lead leg should be straight at the knee as your body passes over it. This may cause some exaggerated hip move-

ments, which are common in race walkers. But walking at a speed slower than 12:00 minutes per mile shouldn't require any exaggerated or bizarre hip movements.

- If you are walking over hilly or mountainous terrain and you try to finish at the point where you started, you will find that you expend about the same amount of energy and get the same amount of exercise as a person who walks on a flat surface. The extra energy that it takes to go uphill is neutralized by the lesser energy required to go downhill.

When you have finished your walk, do not stop suddenly and sit down. Rather, cool down for about five minutes. This phase involves slow, easy walking and some further mild stretching exercises.

Note: The arm action required by fast walking gives the upper body more of a workout than it receives with jogging. Some people have successfully used hand weights weighing two to three pounds in each hand to augment the endurance benefits. I do not recommend heavier weights because of the possibility of injury to the elbows, shoulders, or back.

An Explanation of the Fast Walking Charts

The charts on pages 77–78 are divided into six major columns. The "Week" column refers to the number of weeks you have been following the program. The "Speed/mph" is the number of miles per hour at which you walk. "Min/mi" is the minutes it takes you to go one mile. "Time/distance" is a double column, with the first number referring to the length of time you walk in one workout, and the second to the total distance you walk in that time.

"Freq/week" is also a double column, with the first set of entries indicating the number of times you have to work out per week at this level to achieve a level of athletic fitness. The second line column is the number of times you have to work out per week to achieve a level of fitness that will ensure you good health and maximum longevity.

The final column indicates the recommended level of antioxidants you should be taking to protect yourself from excess free radicals during exercise. As indicated on the charts, you should follow "Recommendation #1" for supplements (described earlier in this chapter) when working out at the health-longevity fitness levels. You should follow "Recommendation #2" for supplements if you work out at the athletic fitness levels.

This program can be used on a motor-driven treadmill if you follow the speed and time guidelines indicated on the charts. Or you can pursue the regimen on a measured track or over a measured distance. You will need an accurate clock or watch with a second hand or digital seconds indicator to help you monitor your time and distance. Each week, your goal is to reach the suggested time, distance, or speed by the end of the week. If your goal is only to achieve a health and longevity level of fitness, there is no need to work beyond the levels indicated for the fifth week for those under fifty years of age, or beyond the eleventh week for those fifty years and older.

Design Your Own Walking Program: An Explanation of the Fast Walking Fitness Points Chart

For those walkers who want to design their own endurance programs, the point charts on pages 80–85 should be of help. For example, you may want to walk a relatively long distance one day in the week and shorter distances on other days. But try to work out at least three days a week, and spread the fitness points you earn as evenly as possible throughout the week. It can be extremely detrimental to your health—and cause the generation of many extra free radicals—if you try to concentrate a great deal of heavy exercise into one day each week.

Remember, your goal should be to achieve at least fifteen points per week if you want to enjoy adequate health and longevity

The Progressive Fast Walking Program
(Under Fifty Years of Age)

Week	Speed/mph	Min/mi	Time/distance	Freq/week: Athletic	Health/ Longevity	Antiox. Suppl.
1	3.0	20:00	20:00 1 mi.	4-5X	3X	Rec. #1
2	3.25	18:45	22:30 1.25	4-5X	3X	
3	3.50	17:30	25:00 1.50	4-5X	3X	
4	3.75	16:15	27:30 1.75	4-5X	3X	
5	4.0	15:00	30:00 2.0	4-5X	3X*	
6	4.25	14:15	28:30 2.0	4-5X		
7	4.50	13:30	27:00 2.0	4-5X		
8	4.75	12:45	25:30 2.0	4-5X		
9	5.0	12:00	24:00 2.0	3-4X		
10	5.0	12:00	27:00 2.25	3-4X		Rec. #2
11	5.0	12:00	30:00 2.50	3-4X		
12	5.0	12:00	33:00 2.75	3-4X		
13	5.0	12:00	36:00 3.0	3-4X		

*Those exercisers wanting only to reach the health and longevity level of fitness need not exceed the week 5 level.

The Progressive Fast Walking Program
(Fifty Years of Age and Older)

Week	Speed/mph	Min/mi	Time/distance	Freq/week: Athletic	Health/ Longevity	Antiox. Suppl.
1	2.50	25:00	25:00 1 mi.	4-5X	3X	Rec. #1
2	2.75	22:30	22:30 1.00	4-5X	3X	
3	3.0	20:00	20:00 1.00	4-5X	3X	
4	3.0	20:00	20:00 1.00	4-5X	3X	
5	3.25	18:45	22:30 1.25	4-5X	3X	
6	3.25	18:45	22:30 1.25	4-5X	3X	
7	3.50	17:30	25:00 1.50	4-5X	3X	
8	3.50	17:30	25:00 1.50	4-5X	3X	
9	3.75	16:15	27:30 1.75	4-5X	3X	
10	3.75	16:15	27:30 1.75	4-5X	3X	
11	4.0	15:00	30:00 2.00	4-5X	3X*	
12	4.0	15:00	37:30 2.5	4-5X		Rec. #2
13	4.25	14:15	28:30 2.0	4-5X		
14	4.25	14:15	35:30 2.5	4-5X		
15	4.50	13:30	27:00 2.0	4-5X		
16	4.50	13:30	33:45 2.5	4-5X		

*Those exercisers wanting only to reach the health and longevity level of fitness need not exceed the week 11 level.

fitness, or to achieve at least thirty-five points per week if you want to move into a higher aerobic fitness range. You do not need to exceed fifty per week unless you are a competitive athlete.

You will note that fitness point values are given for distances and walking speeds from one mile to ten miles. The first column on the far left gives the time you spend walking in hours, minutes, and seconds. The second column indicates the point value you earn for that performance. The third column indicates whether you should follow Recommendation #1 or #2 for taking antioxidants. All the guidelines about technique and the need to seek medical advice apply here, as with the regular progressive program.

The Progressive Treadmill Walking Exercise Program

The following exercise program has been designed especially for use on a treadmill. The rate of increase in the exercise intensity has been keyed to those who are fifty years old or older. Adults in this age group should begin at week one in the program. Younger exercisers may start at a later week in the program if they find they can easily handle a higher level of exercise intensity. Whatever your age, you should only walk, not jog or run, while on this program.

As before, medical clearance should be obtained before you start, particularly if you expect to go further than four weeks into the program. Also, if you have a history of medical problems—especially heart disease—medical clearance is mandatory.

There is no need to go beyond the twelfth week if your goal is limited to health and longevity fitness. If your objective is athletic fitness, continue through week 16 and maintain that level.

The "Week" column indicates the number of weeks you have pursued the program. You should try to reach the speed and time indicated on the same line as the designated week. The "Speed" column is the pace at which you walk in miles per hour. The

The Fast Walking Points
1.0 Mile

Time	Point Value	Antioxidant Suppl. Rec.
over 20:00	0	Rec. #1
20:00-15:01	1.0	
15:00-14:16	2.0	
14:15-13:31	2.75	
13:30-12:46	3.5	
12:45-12:01	4.25	

12:00-11:16	5.0	Rec. #2
11:15-10:31	5.75	
10:30 or less	6.50	

1.5 Miles

Time	Point Value	Antioxidant Suppl. Rec.
over 30:00	0.5	Rec. #1
30:00-22:31	2.0	
22:30-21:26	3.5	
21:25-20:16	4.60	
20:15-18:56	5.75	
18:55-18:01	6.90	

18:00-16:56	8.0	Rec. #2
16:55-15:46	9.0	
15:45 or less	10.25	

2.0 Miles

Time	Point Value	Antioxidant Suppl. Rec.
over 40:00	1.0	Rec. #1
40:00-30:01	3.0	
30:00-28:31	5.0	
28:30-27:01	6.5	
27:00-25:31	8.0	
25:30-24:01	9.5	

24:00-22:31	11.0	Rec. #2
22:30-21:01	12.5	
21:00 or less	14.0	

2.5 Miles

Time	Point Value	Antioxidant Suppl. Rec.
over 50:00	1.5	Rec. #1
50:00-37:31	4.5	
37:30-35:36	6.5	
35:35-33:46	8.40	
33:45-31:51	10.0	
31:50-30:01	12.0	

30:00-28:11	14.0	Rec. #2
28:10-26:16	16.0	
26:15 or less	18.0	

3.0 Miles

Time	Point Value	Antioxidant Suppl. Rec.
over 60:00	2.0	Rec. #1
60:00-45:01	5.0	
45:00-42:46	8.0	
42:45-40:31	10.25	
40:30-38:16	12.5	

Time	Point Value	Antioxidant Suppl. Rec.
38:15-36:01	14.75	Rec. #2
36:00-33:46	17.0	
33:45-31:31	19.25	
31:30 or less	21.5	

4.0 Miles

Time	Point Value	Antioxidant Suppl. Rec.
over 1:20:00	3.0	Rec. #1
1:20:00-60:01	7.0	
60:00-56:46	11.0	
56:45-54:01	14.0	

Time	Point Value	Antioxidant Suppl. Rec.
54:00-51:01	17.0	Rec. #2
51:00-48:01	20.0	
48:00-45:01	23.0	
45:00-42:01	26.0	
42:00 or less	29.0	

5.0 Miles

Time	Point Value	Antioxidant Suppl. Rec.
over 1:40:00	4.0	Rec. #1
1:40:00-1:15:01	9.0	
1:15:00-1:10:16	14.0	

Time	Point Value	Antioxidant Suppl. Rec.
1:10:15-1:07:31	17.75	Rec. #2
1:07:30-1:03:46	21.50	
1:03:45-60:01	25.25	
60:00-56:16	29.0	
56:15-52:31	32.75	
52:30 or less	36.5	

6.0 Miles

Time	Point Value	Antioxidant Suppl. Rec.
over 2:00:00	5.0	Rec. #1
2:00:00-1:30:01	11.0	

Time	Point Value	Antioxidant Suppl. Rec.
1:30:00-1:25:31	12.0	Rec. #2
1:25:30-1:21:01	21.5	
1:21:00-1:16:31	26.0	
1:16:30-1:12:01	30.5	
1:12:00-1:07:31	35.0	
1:07:30-1:03:01	39.5	
1:03:00 or less	44.0	

7.0 Miles

Time	Point Value	Antioxidant Suppl. Rec.
over 2:20:00	6.0	Rec. #1

Time	Point Value	Antioxidant Suppl. Rec.
2:20:00-1:45:01	13.0	Rec. #2
1:45:00-1:39:46	20.0	
1:39:45-1:34:31	25.25	
1:34:30-1:29:16	30.50	
1:29:15-1:24:01	35.75	
1:24:00-1:18:46	41.0	
1:18:45-1:13.31	46.25	
1:13:30 or less	51.5	

8.0 Miles

Time	Point Value	Antioxidant Suppl. Rec.
over 2:40:00	7.0	Rec. #2
2:40:00-2:00:01	15.0	
2:00:00-1:54:01	23.0	
1:54:00-1:48:01	29.0	
1:48:00-1:42:01	35.0	
1:42:00-1:36:01	41.0	
1:36:00-1:30:01	47.0	
1:30:00-1:24:01	53.0	
1:24:00 or less	59.0	

9.0 Miles

Time	Point Value	Antioxidant Suppl. Rec.
over 3:00:00	8.0	*Rec. #2*
3:00:00-2:15:01	17.0	
2:15:00-2:08:16	26.0	
2:08:15-2:01:31	32.75	
2:01:30-1:54:46	39.15	
1:54:45-1:48:01	46.25	
1:48:00-1:41:16	53.0	
1:41:15-1:34:31	59.75	
1:34:30 or less	66.5	

10.0 Miles

Time	Point Value	Antioxidant Suppl. Rec.
over 3:20:00	9.0	*Rec. #2*
3:20:00-2:30:01	19.0	
2:30:00-2:22:31	29.0	
2:22:30-2:15:01	36.5	
2:15:00-2:07:31	44.0	
2:07:30-2:00:01	51.5	
2:00:00-1:52:31	59.0	
1:52:30-1:45:01	66.5	
1:45:00 or less	74.0	

"Incline" column is the incline setting on which you place the treadmill to simulate walking on flat ground or up a hill. The "Time" is in minutes and seconds. The "Freq/Week" column has two parts: the "Athletic Fitness" line for those looking for a higher level of fitness, and the "Health/Longevity" line. The "Antioxidant" line shows whether you should be taking antioxidant supplements according to Recommendation #1 or Recommendation #2, as described earlier in this chapter.

Five Sample Health and Longevity Fitness Programs

The five programs on pages 88–89—regular walking, fast walking, walking-running, cycling, and swimming—are designed for use by those who want to maintain a fitness level that will enable them to maximize their potential for good health and longevity. Each of the programs is set up so as to allow the participant to earn at least fifteen fitness points, the minimum to achieve this level of fitness.

An assumption underlying these programs is that you have already reached a level of fitness that will allow you to cover the distances in the allotted times and in the frequencies per week that are indicated. If you are now a totally sedentary person or otherwise find yourself unable to perform at these recommended levels, you should use one of the previous progressive exercise programs to increase your level of fitness to the required health and longevity level.

I recommend that you exercise at least three times per week, but it is possible to achieve health and longevity fitness—and fifteen points per week—by exercising either in longer or faster sessions two times per week.

As with any other new exercise program, you should undergo a thorough medical exam by a qualified physician before embarking on any of these regimens.

The Progressive Treadmill Exercise Walking Charts
(all ages)

Week	Speed (MPH)	Incline	Time	Freq/week: Athletic	Health/ Longevity	Antiox. Suppl.
1	2.5-3.0	Flat	20:00	5X	3X	*Rec. #1*
2	2.5-3.0	Flat	20:00	5X	3X	
3	3.0	Flat	22:30	5X	3X	
4	3.0	Flat	25:00	5X	3X	
5	3.25	Flat	22:30	5X	3X	
6	3.25	Flat	25:00	5X	3X	
7	3.50	Flat	22:30	5X	3X	
8	3.50	Flat	25:00	5X	3X	
9	3.75	Flat	25:00	5X	3X	
10	3.75	Flat	27:30	5X	3X	
11	4.0	Flat	27:30	5X	3X	
12	4.0	Flat	30:00	5X	3X*	

(under 50 years of age)

Week	Speed (MPH)	Incline	Time	Freq/week: Athletic	Health/ Longevity	Antiox. Suppl.
13	4.0, or	Flat	30:00	5X		*Rec. #2*
	4.0	2.5%	30:00	4X		
14	4.0, or	Flat	35:00	5X		
	4.0	2.5%	35:00	4X		
15	4.0, or	Flat	40:00	5X		
	4.0	5.0%	40:00	4X		
16	4.0, or	Flat	45:00	5X		
	4.0	5.0%	45:00	4X		

*Those exercisers wanting only to reach the health and longevity level of fitness need not exceed the week 12 level.

Sample Health and Longevity Walking Program

Cover 2.0 miles in less than 40:00 minutes, five times per week
= 15 Fitness points
Cover 2.0 miles in less than 35:00 minutes, four times per week
= 16 points
Cover 2.0 miles in less than 30:00 minutes, three times per week
= 15 points
Cover 3.0 miles in less than 45:00 minutes, two times per week
= 16 points

Sample Health and Longevity Aerobic Walking Program

Cover 1.5 miles in less than 20:15 minutes, three times per week
= 17.25 points
Cover 2.0 miles in less than 28:30 minutes, three times per week
= 19.5 points
Cover 2.0 miles in less than 27:00 minutes, two times per week
= 16.0 points
Cover 2.5 miles in less than 35:35 minutes, two times per week
= 16.8 points

Sample Health and Longevity Walking-Running Program

Cover 1.5 miles in less than 18:00 minutes, three times per week
= 15 points
Cover 2.0 miles in less than 20:00 minutes, two times per week
= 18 points
Cover 2.5 miles in less than 30:00 minutes, two times per week
= 18 points

Sample Health and Longevity Cycling Program

Cover 5.0 miles in less than 30:00 minutes, five times per week
= 17.5 points

Cover 5.0 miles in less than 20:00 minutes, three times per week
= 18.0 points
Cover 6.0 miles in less than 36:00 minutes, four times per week
= 18.0 points
Cover 7.0 miles in less than 42:00 minutes, three times per week
= 16.5 points

Sample Health and Longevity Swimming Program

Cover 500 yards in less than 16:40 minutes, five times per week
= 15.6 points
Cover 500 yards in less than 12:30 minutes, four times per week
= 16.7 points
Cover 600 yards in less than 15:00 minutes, three times per week
= 15.0 points
Cover 800 yards in less than 20:00 minutes, two times per week
= 15.3 points

Five Sample Athletic Fitness Programs

The following five programs—regular walking, fast walking, walking-running, cycling, and swimming—are designed for use by those who want to maintain a fitness level that will enable them to maximize their athletic fitness—and enjoy the enhanced sense of well-being and other benefits associated with this higher level of conditioning. Each of the programs is set up to allow the participant to earn at least thirty-five fitness points, the minimum to achieve this level of fitness. You will note that none of the programs exceeds fifty fitness points per week—the upper threshold, beyond which the body's production of free radicals increases significantly during exercise.

An assumption underlying these programs is that you have already reached a level of fitness that will allow you to cover the distances in the allotted times and in the frequencies per week that are indicated. If you are now a totally sedentary person or otherwise find yourself unable to perform at these recommended levels, you should use one of the previous progressive exercise programs to increase your level of fitness to the required athletic fitness level.

I recommend that you exercise at least four times per week, but it is possible to achieve athletic fitness—and thirty-five points per week—by exercising either in longer or faster sessions three times per week.

As with any other new exercise program, you should undergo a thorough medical exam by a qualified physician before embarking on any of these regimens.

Sample Athletic Fitness Walking Program

Cover 3.0 miles in less than 45:00 minutes, five times per week = 40 points

Cover 3.5 miles in less than 52:30 minutes, four times per week = 38 points

Cover 4.0 miles in less than 60:00 minutes, four times per week = 44 points

Cover 5.0 miles in less than 75:00 minutes, three times per week = 42 points

Sample Athletic Fitness Aerobic Walking Program

Cover 2.0 miles in less than 27:00 minutes, five times per week = 40 points

Cover 2.0 miles in less than 25:30 minutes, four times per week = 38 points

Cover 2.5 miles in less than 35:35 minutes, four times per week = 34 points

Cover 3.0 miles in less than 40:30 minutes, three times per week = 37.5 points

Cover 4.0 miles in less than 56:45 minutes, three times per week = 42 points

Sample Athletic Fitness Walking-Running Program

Cover 2.0 miles in less than 24:00 minutes, five times per week = 35.0 points

Cover 2.0 miles in less than 20:00 minutes, four times per week
= 36.0 points

Cover 2.5 miles in less than 30:00 minutes, four times per week
= 36.0 points

Cover 2.5 miles in less than 25:00 minutes, three times per week
= 34.5 points

Cover 3.0 miles in less than 30:00 minutes, three times per week
= 42 Points

Sample Athletic Fitness Cycling Program

Cover 5.0 miles in less than 20:00 minutes, six times per week
= 36 points

Cover 6.0 miles in less than 24:00 minutes, five times per week
= 37.5 points

Cover 7.0 miles in less than 28:00 minutes, four times per week
= 36.0 points

Cover 8.0 miles in less than 32:00 minutes, four times per week
= 42.0 points

Cover 10.0 miles in less than 40:00 minutes, three times per week
= 40.5 points

Sample Athletic Fitness Swimming Program

Cover 750 yards in less than 18:45 minutes, five times per week
= 35.0 points

Cover 800 yards in less than 20:00 minutes, five times per week
= 38.3 points

Cover 900 yards in less than 22:30 minutes, four times per week
= 36.0 points

Cover 1,000 yards in less than 33:20 minutes, four times per week
= 33.0 points

Cover 1,200 yards in less than 30 minutes, three times per week
= 39.0 points

5

The Strength Training Program

uring the course of one recent workday, I encountered these complaints from patients and friends:

- A fifty-eight-year-old nutritionist, who had just returned from a trip to the mountains where she had done a great deal of downhill walking, said that the front of her legs and feet were quite sore.
- A thirty-four-year-old salesman reported that his upper thighs, groin, and buttocks were tight and painful as a result of sprinting he had done a couple of days before as part of an interval training routine. (His regimen involved alternating fast anaerobic, sprint-type running with slower, long distance work.)
- A forty-five-year-old plant manager could hardly lift his arms above his head after doing some push-ups for the first time since college.
- A sixty-eight-year-old grandmother experienced excruciating soreness one day after she had spent an hour rearranging the furniture in her home.
- A twenty-nine-year-old attorney, who had just embarked on a weight-training program, said she was feeling so many aches and pains in so many places that she questioned whether or not her workouts were really good for her health.

In fact, such unaccustomed muscular exertion may *not* be good for your health, particularly when your body is unprepared to deal with unusual physical tasks. The problem is similar to what we confronted with excessive, high-intensity endurance exercise: Engaging in short bursts of heavy work or physical movements that exceed the demands you normally place on your muscles and joints may cause a multitude of problems, including the release of free radicals. The radicals associated with these overworked muscles may, in turn, place you at additional risk for a cornucopia of diseases.

What Is the Evidence for Free Radical Damage During Strength Work?

Although most of the studies relating to free radical damage, antioxidants, and exercise have focused on high-intensity endurance exercise, a number of suggestive findings and observations have also emerged to implicate excessive strength work.

Here is the basic premise, stated well by researcher Vishwa N. Singh in 1992 symposium proceedings published by the American Institute of Nutrition:

> There is good scientific rationale for strenuous exercise potentially leading to excessive free radical production. . . . Several lines of evidence suggest that among other factors, free radical damage might contribute to the etiology of chronic diseases, such as cancer, cardiovascular disease, cataract, etc.

A British study supporting that premise involved patients aged twenty-five to seventy-eight with rheumatoid arthritis and knee joint damage. Participants, most of whom were women, did an isometric quadriceps contraction exercise for two minutes. In other words, they contracted the muscles in their thighs as intensely as they could for the allotted time period. Then, fluids in the knees were measured to determine if there was any sign of TBARS, the residue that indicates the presence of free radicals.

Researcher Peter Merry and associates at the London Hospital Medical College came to this conclusion in a 1991 report in the *American Journal of Clinical Nutrition:* "The study shows that isometric quadriceps exercise of inflamed joints produces significantly elevated levels of TBARS in synovial fluid, when compared to pre-exercise values, at all the post-exercise time points studied."

Free radical damage, then, may result simply from the act of putting extra pressure on joints and muscles during exercise. In addition, the "ischemia-reperfusion" phenomenon discussed in the previous chapter may come into play. That is, when you exercise one set of muscles intensely, you may temporarily divert blood away from surrounding joints and tissues ("ischemia," or the deprivation of oxygen and blood). When the heavy activity ceases, the blood then rushes back into the deprived tissues ("reperfusion"). The process is known to trigger the output of free radicals and increase the potential for damage to the body's cells.

Ischemia-reperfusion may occur during maximal, strength-oriented exercise such as competitive rowing, where you use more oxygen than you take in as you test the stamina of your shoulders, arms, back, and legs. At a symposium entitled "Antioxidants and the Elite Athlete," sponsored at meetings of the American College of Sports Medicine in Dallas in May 1992, one participant, Mitchell Kanter, warned about the impact on the body of this exhaustive rowing-type activity. He said that "super-maximal activity can potentially produce a transient ischemia/reperfusion situation known in other instances to greatly exacerbate free radicals."

Kanter also suggested that "it may well be that with this type of exercise you would be in more need of antioxidant protection." That is precisely the angle that other scientists have explored in evaluating possible injury to the body through strength-centered exercise.

At the Australian Institute of Sport in Canberra, Australia, for instance, Dr. Ian Gilliam helped conduct a study of several groups of top athletes, including cross-country skiers and triath-letes, who must rely on muscle strength and stamina as well as aerobic endurance. He put one group of the athletes on a placebo, and another group on daily antioxidant supplements in these amounts: 1,000 IU of vitamin E and 1,000 mg of vitamin C. The athletes taking the supplements showed a 25 percent reduction in tissue damage, as measured by decreases in enzymes associated with overtraining.

Gilliam and his colleagues concluded: "This suggests that the membranes in muscle, including cardiac muscle, sustained less oxidative damage when the athletes were supplemented. The de-crease in lactate dehydrogenase [an enzyme linked to overtraining] suggests that the red blood cells may also sustain less damage."

In a related finding, Gilliam and his group tested the effect of the exhaustive exercise on the testosterone levels of the athletes. Usually, this hormone will fall markedly during heavy training. But in the supplemented athletes, he discovered that the testoster-one levels were much higher than in those on the placebo. "This is a very interesting finding because it means that with antioxidant supplementation, muscles can recover and regenerate more quickly after exercise," he concluded.

Mountain climbers, who in many ways are the quintessential strength athletes, must have highly conditioned muscles through-out their bodies to perform well in their sport. A study by I. Simon-Schnass and H. Pabst, reported in 1988 in the *International Journal of Vitamin and Nutrition Research*, confirmed that this demanding activity, when performed at high altitudes, results in increased pentane production in exhaled breath. Once again, pentane is an indicator of the presence of excess free radicals. But when the climbers took daily 200 mg supplements of synthetic vitamin E for four weeks, their pentane production decreased and their work performance improved.

In a related finding, researchers at the Department of Agricul-ture's Human Nutrition Research Center on Aging at Tufts Univer-sity in Boston found that vitamin E supplements could prevent much of the free radical injuries produced by heavy exercise. Specifically, the researchers gave half of the study participants 800 IU of vitamin E each day for seven days and the other half a placebo. They directed the subjects to run downhill on a treadmill for forty-five minutes. As you know, putting that kind of stress on the leg muscles is likely to produce soreness because most people do not walk downhill for such a long period of time.

The results, which were reported in October 1992, revealed that the participants on the placebo experienced significant soreness after the workout. But the people who took the vitamin E experienced much less soreness and produced fewer chemicals in the body associated with inflammation and overuse of muscles.

Is there any way that you can tell whether you are going too far with your training? Are there any telltale signs or signals that can alert you to a need to adjust or cut back on your routine? Here are a few guidelines that I find helpful in avoiding the "overtraining syndrome."

Avoiding the Overtraining Syndrome

The medical perspective on overtraining came through clearly at a panel discussion during the 39th Annual Meeting of the American College of Sports Medicine in Dallas on May 27, 1992. There Ian Gilliam of the Phillip Institute of Technology in Australia observed:

The overtraining syndrome is characterized by a whole range of clinical indices. Some of the most pronounced are: falls in testosterone; rise in cortisol [a steroid hormone secreted by the adrenal glands]; falls in the testosterone/cortisol ratio; hemolysis [destruction of red blood cells]; red blood cell sports anemias; reduced performances; elevated cytoplasmic enzymes [secretions from the non-nucleus part of a cell]; and in females: amenorrheas [a failure to menstruate].

In more practical terms, you may want to pay attention to this checklist that signals an overtrained—and probably a high free radical—state. These points were compiled by Dr. Neil F. Gordon at the Cooper Institute for Aerobics Research:

- Changes in your sleep patterns, especially insomnia
- Longer healing period for minor cuts and scratches
- Fall in blood pressure and dizziness when getting up from a prone or seated position
- Gastrointestinal disturbances, especially diarrhea
- Gradual loss of weight in the absence of dieting or increased physical activity
- Greater than usual increase in heart rate during a standard exercise session
- Leaden or sluggish feeling in legs during exercise
- Impaired mental acuity and performance or inability to concentrate
- Inability to complete routine exercise training sessions that were no special challenge previously
- Increase in resting heart rate (recorded early in the morning) by more than ten beats per minute
- Excessive thirst and fluid consumption at night
- Greater susceptibility to infections, allergies, headaches, and injuries
- Lethargy, listlessness, and tiredness
- Loss of appetite
- General loss of enthusiasm, drive, and motivation (In athletes or other people who usually derive joy from exercise, this lack of interest and incentive extends to sports.)
- Loss of libido or interest in sex
- Irregular or no menstruation in premenopausal women
- Muscle and joint pains
- Sluggishness that persists for more than twenty-four hours after a workout
- Swelling of the lymph nodes

These points are just signals that you may be involved in over-training. If you observe one or more of them in your life and you suspect too much exercise may be the cause, try cutting back on your workouts and see what happens. If you find that the problem is corrected, you may very well have not only improved your current well-being, but also reduced your exposure to destructive free radicals.

The Strength Training Dilemma

It is clear that a mounting body of evidence is alerting us to the danger of placing heavy demands on our muscles and triggering the production of excess free radicals, which place us at greater risk of disease. Yet other lines of research have demonstrated definitively that strength training is a must for good health, especially as we grow older.

For one thing, as I showed in my book *Preventing Osteoporosis* (Bantam, 1989), we begin to lose our bone mass at the rate of about 1 percent per year after our mid-thirties, and that rate escalates for women who have passed the menopause. An essential step in preventing bone loss is regular, weight-bearing exercise such as the strength training discussed in this chapter.

Furthermore, it is essential for all adults, regardless of their risk of osteoporosis, to pursue a regular strength training program in order to preserve their basic ability to function physically. One of the great tragedies of our society is that we are increasing the longevity of men and women without teaching them to retain their ability to operate independently. The average person spends approximately one decade with some form of incapacity before death—and for the most part, impairments of physical ability can be prevented, or at least minimized, with a regular strength program.

So this is the dilemma: If you engage in excessive strength work or place demands on your body that your muscles are unprepared to meet, you will generate excess free radicals that will put you at an increased risk for serious diseases. But if you fail to do strength work, you will most likely find your functional capacity decreasing to unacceptable levels as you age. What is the answer?

The answer may be found in three areas:

1. You must be sure that you select foods that are high in the antioxidants vitamin E, vitamin C, and beta carotene. Chapter 7 will show you how.

2. You should incorporate a personalized antioxidant supplement program into your daily routine, as described in chapter 6. Studies described above show how important it is to utilize supplements when you are doing strength work.

3. You should embark on a progressive, lower-intensity flexibility and strength program that will prepare your body for most of the expected and unexpected physical demands you will face in the course of a typical day. The following regimen has been designed to build up your flexibility and strength so that you can enhance your tolerance for the high demands that will inevitably be placed on your muscles and bones during your daily activities.

A Strength Program for the Antioxidant Revolution

The following strength program is divided into three parts—flexibility exercise, calisthenics, and exercises using free weights and machines. Each program has been designed to increase your flexibility and strength *gradually*. Remember, very intense and demanding strength work—such as exercising until you are exhausted or trying to lift as much weight as you can in one or two repetitions—may increase not only your muscle size and strength, but also your body's output of free radicals. So stick to the program as outlined here, and move from one level of intensity to the next systematically and gradually. Do not give in to the temptation to "prove how strong you are."

The exercises are taken from *The Strength Connection* and from other manuals produced and published at the Cooper Institute for Aerobics Research in Dallas. Instructions are included with each individual exercise, but first you should design an overall flexibility-strength plan for yourself. Here are some suggestions for how to do that.

Do flexibility and strength exercises at least two to three times a week, preferably on your "off" days, when you aren't doing your endurance work. Plan on devoting twenty to thirty minutes to each flexibility-strength session. One-third of your time should be spent doing the flexibility exercises, and two-thirds should be spent doing strength work.

Follow the instructions for each exercise carefully to ensure that you stay on a progressive and gradual program. Again, do not try to push yourself too fast, or you may suffer injury or discomfort from sore muscles.

You should try to exercise each of the main muscle groups (e.g. chest, arms, abdomen, thighs) during each workout—but you do not have to do the same exercise every session. Feel free to vary your routine; that is a good way to build overall strength and flexibility fitness *and* to keep your interest high. Similar—and sometimes overlapping—exercises have been included to enable you to inject some variety and still get an adequate workout for all parts of your body.

When you do your strength work, you can concentrate on calisthenics during one session and weights the next; or you can mix calisthenics and weights during the same session. For example, you might do pushups one day for your chest and arms and bench presses the next time for the same muscle groups. If you do shift around from one type of exercise to another, be sure that you keep track of the repetitions, sets, or resistance levels that you use so that you can proceed systematically with your strength development. The main idea is to have fun as you experience a wide-ranging and open-ended workout that will produce the results you want—without overdoing it and stimulating harmful free radicals.

Now, with this plan in mind, plug in the exercises that appeal to you for each major muscle group.

Flexibility Exercises

These stretching movements have three main purposes. First, they will help you warm your body up for more demanding activity. Consequently, they should be done before the actual strength work. Second, the stretches will enable you to improve the range of motion of different parts of your body so that you will be able to perform the strength movements more easily. Third, the more flexible you are and the better warmed up, the less likely you will be to experience injuries during the strength work.

There are two phases to a stretch: active and static. Take the active or moving phase of each stretch to an easy point of tension. Do not try to force your body to go beyond a slight feeling of tension, or you may experience injury. After you reach your easy tension point, start the static part of the stretch this way: Hold your position so that the tension on the muscles you are exercising lasts for ten to thirty seconds. Then slowly release the tension and return to your starting position.

Perform all the stretching exercises slowly and with your body under full control. Never stretch to the point of pain, and do not bounce while your muscles are fully stretched. Bouncing can cause injury.

Proper breathing is also important. Inhale before the stretch, and exhale during the active phase of the stretch.

Note: These flexibility movements can be used as part of your warmup for your endurance workouts, as well as for the strength program.

1: Chest Stretch

Stand facing a door frame. Place your hands on each side of the doorway at shoulder level, and walk through the door until you begin to feel a slight tension across your chest muscle. Hold for 20 seconds.

2: Upper Back Stretch

Stand facing one side of a door frame or an upright pole. Place your hands on the frame or pole at about shoulder level. Lean back and hunch your shoulders slightly so that you feel the muscles of your upper back stretch apart. Hold for 20 seconds.

3: Side Bends

Stand with your feet shoulder-width apart. Extend your right arm overhead, and place your left hand on your left hip. Bend to the left (the side opposite the raised arm). Hold for 10–20 seconds. You should feel tension along your right side. Repeat the movement with the other side of your body.

4: Standing Thigh Stretch

Stand facing a wall, and brace yourself against the wall with your right hand. Raise your right foot up toward your buttocks and grasp it along the instep or ankle with your left hand. *Gently* pull your right foot up closer to your buttocks so that you begin to feel tension in your right thigh (but *not* your right knee). Hold for 10–20 seconds. Then, repeat the movement with the other leg.

5: Achilles Tendon and Calf Stretch

Stand facing a wall, with your right foot in advance of your left. Your right heel should be about 12–18 inches in front of your left toe. Lean forward, placing the palms of your hands flat against the wall. Your right knee should be bent, and your left knee almost straight. Slowly move your hips forward until you feel a stretch in the calf of your left leg. Keep both feet and heels flat on the ground. Hold for 15–30 seconds. Then shift the tension to your left Achilles tendon by bending your left knee slightly while keeping your left heel on the ground. Repeat these movements with the other leg.

Fig. 1   Fig. 2

6: Standing Hamstring Stretch

Stand with your feet wide apart in a straddle position with toes pointing forward. Turn your left foot out, and bend the left knee, keeping your body weight over the left leg with your chest parallel to the floor. Keep your right leg straight, and place your hands behind your back (Fig. 1). While holding the chest parallel to the floor, slowly straighten the left knee and hold for 15–30 seconds (Fig. 2). Repeat with right leg. The stretch should be felt in the back of your forward leg, and in the hamstring tendons and muscles (located along the back of your upper leg).

7: Hamstring Stretch With One Leg Up (Optional)

Stand facing a platform, railing, or other elevated surface which is about waist-high. Raise your left leg in front of your body, and rest your left heel on the platform. Bend forward at the hips and lean your upper torso toward the raised leg. Hold for 15–30 seconds. Repeat with the right leg. The stretch should be felt in the hamstring of the raised leg and in the lower back.

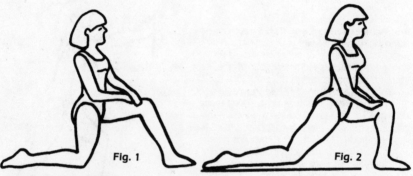

Fig. 1　　Fig. 2

8: Hip Flexor Stretch

Place your right knee on the floor behind your left foot (Fig. 1). Move your left foot forward, with your foot flat on the ground and your left toe pointed forward, so that the heel of your left foot is slightly in advance of your left knee.

Without changing the position of your right knee or left foot, push your hips forward and downward (Fig. 2). You should feel tension in the front part of your right hip and also in your groin and hamstrings (the muscles running down the back of your upper leg). Hold this position for 15–30 seconds.

Repeat the motion with the other leg.

9: Sitting Hamstring Stretch (Optional)

Sit on the floor with both legs stretched out in front of your body. Keep your knees straight and your feet approximately a yard apart. Bend forward at the hips, lean your upper torso toward the left leg, and grasp your left ankle with both hands. Hold for 15–30 seconds. Repeat with your right leg. The stretch should be felt in the hamstrings and lower back.

10: Gluteal Stretch

Sit on the floor with both your legs straight out in front of your body. Your legs should be about a yard apart. Slip both your arms around your left ankle and knee, and raise your left leg up toward your chest. You should feel tension in your gluteal region—your outer thigh and buttocks. Hold this position for 15–30 seconds. Then repeat the motion with your other leg.

11: Groin Stretch

Sit on the floor with the soles of your feet together. Place your hands around your feet to stabilize their position, and pull the heels of your feet in toward your groin (Fig. 1). Lean forward, with your upper arms or elbows pushing against the inside of each knee, so that you push your knees toward the floor (Fig. 2). You should feel tension in your groin and lower back. Hold this position for 15–30 seconds.

Fig. 1

Fig. 2

12: Lying Thigh (or quadriceps) Stretch

Lie on your stomach, with your head down on the floor. With your right hand, grasp your right foot and pull it back *gently* toward your buttocks. Tension should be felt in your right quadriceps muscle (along the front of the right thigh)—but *not* in your right knee. Hold this position for 15–30 seconds. Then repeat the motion with the left leg.

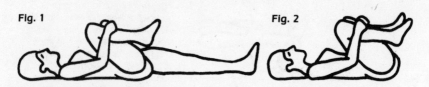

Fig. 1   Fig. 2

 # 13: Lower Back Stretch
Lie on your back, clasp your hands around your right knee, and pull the knee up toward your chest (Fig. 1). Round your lower back as you pull so that you feel the stretch in your lower back. Hold for about 20–30 seconds. Repeat the same motion with your other leg. Finally, clasp your hands around both knees and pull them both up toward your chest (Fig. 2). Hold for about 20–30 seconds. Again, the stretch should be felt in your lower back.

Calisthenics Exercises

Calisthenic exercises allow you to use your body's weight and the force of gravity as resistance to build up your muscle power. For some muscle groups such as your abdominal muscles, calisthenics may be the exercise of choice. (Sit-ups and "crunches" are examples of calisthenics that are best for the abdominal area.)

Another advantage of calisthenics is that no equipment is needed. You can do them at home if you lack gym equipment, and you can also do them when you're travelling or away from available exercise equipment. Because of their ready availability, I encourage everyone to include some calisthenics movements in their daily routine.

The number of repetitions of any calisthenics exercise you perform will vary depending on the exercise involved. As your strength increases, you can modify the exercises to increase the resistance, or the work performed by the particular muscle group. For purposes of these exercises, a "repetition" is one complete movement of an exercise. A "set" is a group of repetitions performed sequentially. If you are doing more than one set, as will often be the case on a progressive program, you should rest about twenty to thirty seconds after the first set and then proceed to the next set. Here is an example involving the push-up:

Assume that you can perform one or two sets of eight to twelve repetitions of push-ups. Then, over the course of a couple of weeks, you find that it is easy for you to do three sets of twelve repetitions each. Generally, if you can complete three sets of eight to twelve repetitions, but no more, then the exercise is just enough to help you build strength.

Once you can easily complete three sets of twelve repetitions, consider making the exercise harder. To that end, you might increase the number of repetitions in each set; or you might add a fourth set; or you might elevate your feet so that your arms have more weight to lift.

To build muscle strength, you must follow a mild "overload" principle, which requires that you add repetitions or sets to your routine to make the workout slightly harder. But remember our basic progressive exercise guidelines: Do not push your body far beyond the next level of intensity. If you do, you may become too exhausted or experience soreness or injury—signs that you may be exposing yourself to injury, including free radical damage.

14: Four Types of Push-ups

Each of these exercises will strengthen the muscles of your chest, shoulders, and back of the arms (triceps).

1. Push-away (easiest). Stand about 3–4 feet from a wall and lean up against the wall, with your palms flat against the wall, supporting your weight. To do one push-away, flex your elbows so that your face is nearly touching the wall, and then push yourself away from the wall to a straight-up, standing position. Then allow your body to fall back toward the wall, so that your face is close to the wall again. This constitutes one repetition. For suggestions about how to use this exercise during one exercise session, see the general instructions at the beginning of this section.

2. Bent knee push-up (second easiest). Kneel on the floor, with your hands flat on the floor in front of your body, shoulder-width apart. Rest your weight only on your knees and your hands, and keep your elbows straight and your arms extended. Lower your upper body to the floor so that your chin barely touches the ground. Then push your upper body back up so that your elbows are once again straight, with arms extended. This constitutes one repetition. See the general instructions at the beginning of this section for using this exercise during an exercise session.

3. *Regular push-up.* Rest your weight on your toes and hands, with your back straight and your arms and elbows straight. Your palms should be flat on the ground about one shoulder-width apart. Lower your body to the floor, using only your arms. Then push your body up so that your elbows are straight once more. This constitutes one repetition. See the general instructions at the beginning of this section for using this exercise during a workout.

4. *Advanced push-up, with feet elevated (most difficult).* Place your feet on an elevated surface, such as a chair or stool. Next, place your hands flat on the ground, with your arms and elbows straight and your entire body extending in one straight line from feet to head. Lower your body using only your arms until your chin touches the floor. Then push your body up again to the position where your arms and elbows are extended, with your entire body straight. This constitutes one repetition. See the general instructions at the beginning of this section for using this exercise during a workout.

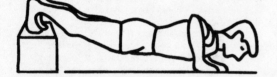

15: Forward Lunges

Stand with your feet shoulder-width apart. Take a giant step forward with your left leg. Keeping your back straight, lower your torso as far as you can toward the floor. Your left leg should remain perpendicular to the ground during this movement so that you avoid flexing your knee and putting undue stress on it. Then, using your left leg, push your body up to the starting position. This constitutes one repetition. Do several repetitions using your left leg, according to the general instructions at the beginning of this section. Then repeat the exercise with the right leg. This exercise will strengthen your thigh muscles.

16: Pull-Ups

Using an overhand grip (palms facing away from your body) on a sturdy, horizontal bar, begin by hanging from the bar with arms and elbows straight. Your feet should not be able to touch the ground. Pull your body up toward the bar until your chin is just above the bar. Then lower your body slowly to the starting position. Exhale when you pull your body up; inhale when you lower it. This constitutes one repetition. For general instructions in using this exercise during a workout, see the information at the beginning of this section.

A variation: If you find you can't do one complete pull-up, you should begin with a lower bar, which allows you to grasp the bar with bent elbows while you're standing on the ground. You can either perform the pull-ups by pulling your body up from this half-way position; or, if you need further help, you can jump off the ground and using this momentum, pull your body the rest of the way up toward the bar.

These exercises will strengthen your biceps (the muscles on the front of your arms), your shoulders, and your back.

17: Calf Raises

Standing erect, with your knees kept straight, raise your weight up on your toes as high as possible. Return to the starting position. If you need to balance yourself, place your hands lightly against a wall or on a bar (but don't allow your hands and arms to give you help raising your body!). This constitutes one repetition. For using this exercise during a workout, see the general instructions at the beginning of this section.

Variations: To make the exercise more challenging, you can do these calf raises on one foot alone, and then on the other. Or you can stand on the edge of a stair-like surface, with your heels hanging over the edge.

These exercises will strengthen your calf muscles.

18: Sit-ups

Lie flat on your back on the floor, with your arms crossed over your chest, your knees bent at about a 45-degree angle, and your feet flat on the floor. Raise your upper body to a sitting position, and then lower your upper body back to the floor. This constitutes one repetition. (If necessary, you can anchor your feet under a bed or have someone hold them down on the ground.) For using this exercise during a workout, see the general instructions at the beginning of this section. This exercise will strengthen your stomach muscles.

Machine and Weight Training Strength Exercises

Resistance training done with free weights and machines provides the most effective and flexible kind of strengthening and muscle-building exercises. The major advantage of weight and machine training over calisthenics is that you can adjust your overload more easily, precisely, and gradually as your strength increases.

To determine the amount of resistance that is appropriate for your current strength, you will have to rely on a process of trial and error. But always begin with a weight that you are certain is too light and work up to find the optimum exercise weight. If you start with weights that are too heavy, you could injure yourself.

Let me mention one point that may seem obvious but that is often overlooked by those with no weight training experience. You will always be able to lift more weight with the larger muscles of your legs than with your arms, chest, or shoulder muscles. So if you find that you can curl a certain amount of weight with your arms, assume that you can lift more with a leg press. Conversely, if you determine you can do a leg press with a certain amount, *do not* assume that you can curl an equal amount. Reduce your trial weight for a curl significantly below the leg press weight.

Here are some general rules of thumb to keep in mind, especially if you have never used weights before:

- Find a resistance or weight with which you can do at least eight but no more than twelve repetitions. Then, perform a minimum of two sets at your selected repetition level for each exercise.
- When you can do sixteen repetitions for three sets, add enough resistance so that you do eight repetitions for three sets. Continue with the resistance until you can perform fifteen to sixteen repetitions. Then repeat the cycle.
- Generally, the heaviest resistance you can do for eight to twelve repetitions represents approximately 60 percent of your maximum lifting ability (the maximum weight that

you can lift one time). This 60-percent-of-maximum principle should provide an ideal guide for developing muscle definition, strength, and even some endurance—without causing injury or triggering the output of free radicals.

19: Leg Press

Set the weight level that is appropriate for you (see general instructions at the beginning of this section). Sit in the apparatus, and place your feet on the foot supports in front of you. Keep your back flat against the seat pad. With your hands, grab the handles on the side of the seat. Extend your legs slowly until your knees are almost straight, but don't "lock" your knees into a straight position. Return the weights slowly to the starting position. This constitutes one repetition. Repeat according to the general workout instructions at the beginning of this section. This exercise will strengthen your thighs.

20: Leg Curl

Set the appropriate weight level. Lie on your stomach on the apparatus with your knees just beyond the edge of the main pad and your heels under the weighted roller pads. Hold the handles loosely. Pulling up against

the roller pads, raise your lower legs up as far as possible toward your buttocks. Slowly allow the roller pads to return to the starting position. This constitutes one repetition. For using this exercise during a regular workout, see the general instructions at the beginning of this section. This exercise will strengthen the backs of your upper legs.

21: Leg Extension

Set the appropriate weight level. Sit in the apparatus with your feet behind the weighted roller pads and your knees just over the front edge of the seat. Fasten the seat belt. Hold the handles loosely. Using the fronts of your ankles, smoothly "kick" the roller pads upward until your legs are fully extended. Then lower the roller pads slowly. Be sure to keep your back against the seat pad. This constitutes one repetition. To use this exercise during a regular workout, see the general instructions at the beginning of this section. This exercise will strengthen the front of your upper legs and your knees.

22: Pull-Downs

Set your appropriate weight level. Grip the bar with your hands wide apart. With your elbows back and a slight forward lean, pull the bar down until it touches the nape of your neck. Slowly allow the bar to return to the starting position. Concentrate on pulling with the muscles in the middle and side of your back. This constitutes one repetition. To use this exercise during a regular workout, see the general instructions at the beginning of this section. This exercise will strengthen your back, biceps, and shoulders.

23: Pull-Downs with Pulleys

Adjust the pulley arms of the Nordic Fitness Chair to the upward vertical position. Grasp the handles with the palms facing forward. Pull the handles downward to shoulder level, and touch your elbows to the side of your chest wall. Pause, and then raise to the original starting position. This constitutes one repetition. To use this exercise during a regular workout, see the general instructions at the beginning of this section. This exercise will strengthen your entire upper arms, your shoulders, and your upper back.

24: Overhead Press

Set the appropriate weight level. Fasten the seat belt. Grip the handles of the bar, and push the bar straight up overhead, keeping your elbows wide apart. Don't arch your back. Slowly lower the bar. This constitutes one repetition. To use this exercise during a regular workout, see the general instructions at the beginning of this section. These movements will strengthen the backs of your upper arms and your shoulders.

25: Triceps Extension

Set your appropriate weight level. Stand facing the apparatus, grasping the bar with both hands, palms down. With your elbows bent and held close to the body, push the bar down to a fully extended position. Allow the bar to rise slowly to the starting position. This constitutes one repetition. To use this exercise during a regular workout, see the general instructions at the beginning of this section. These movements will strengthen the back of your upper arms (your triceps).

26: Abdominal Crunch

Adjust the pulley arm of the Nordic Fitness Chair to the upward vertical position so that you can grasp the handles beside your head. Holding them firmly, draw your head to your knees, keeping your lower back rounded. Contract or tighten your abdominal muscles while performing the exercise. Return to an erect position. This constitutes one repetition. To use this exercise during a regular workout, see the general instructions at the beginning of this section. These movements will strengthen your stomach muscles and also, to some extent, your shoulders and upper arms.

27: Lunge with Dumbbells

Select the appropriate weight level. Stand holding a dumbbell in each hand, with arms hanging down at your sides. Take a giant step forward with your right foot, and lean forward with your back straight. You should be bending your right knee, but keep the lower part of your right leg perpendicular to the ground. Step back to the starting position, standing with the weights at your sides. This constitutes one repetition. To use this exercise during a regular workout, see the general instructions at the beginning of this section. These movements will strengthen your thigh muscles.

28: Bench Press

Select the appropriate weight level. Lie on your back on the bench with your feet flat on the floor. Grasp the barbell with your hands, which should be placed slightly wider than shoulder-width apart. Begin the exercise by holding the barbell so that it's just touching the chest. Keeping your elbows close to your body, raise the barbell straight above your mid-chest level by extending your arms completely. Lower the barbell slowly to your chest. This constitutes one repetition. To use this exercise during a regular workout, see the general instructions at the beginning of this section. These movements will strengthen your chest and arms.

29: Supine Fly

Select the appropriate weight level. Lie on your back on a bench with your feet flat on the floor on either side of the bench. Hold a dumbbell in each hand with palms up, arms bent, and elbows held out to the side as far as is comfortable. Keeping your elbows bent, raise the dumbbells in an arc so that they meet over your chest. Lower the dumbbells slowly to the level of the bench. This constitutes one repetition. To use this exercise during a workout, see the general instructions at the beginning of this section. These movements strengthen your chest and the muscles on the front of your arms and wrists.

30: Pullover

Select the appropriate weight level. Lie on a bench with your head just at the edge. Grip a plate of one dumbbell between your thumbs and index fingers, with arms extended fully above the chest. Slowly lower the weight in a backward arc over and behind your head, bending your elbows slightly. Then raise the weight slowly to the fully extended position. This constitutes one repetition. To use this exercise during a workout, see the general instructions at the beginning of this section. These movements strengthen the backs of your arms and your wrists.

31: Overhead Press

Select the appropriate weight level. This exercise can be performed either standing or seated, or in front of the body or behind the neck. Use a wide grip on the barbell, with palms turned out. Raise the barbell to shoulder level. Extend the arms straight up overhead. Then slowly lower the bar to the shoulder level. This constitutes one repetition. To use this exercise during a regular workout, see the general instructions at the beginning of this section. These movements will strengthen your shoulders and the backs of your upper arms.

32: Side Shoulder Raises

Select the appropriate weight level. Stand with your legs shoulder-width apart. Hold the dumbbells together, arms hanging fully extended at your side. Keeping your elbows slightly bent and your head up, raise the dumbbells up along the side of your body to shoulder height. Then lower them slowly to the starting position. This constitutes one repetition. To use this exercise during a regular workout, see the general instructions at the beginning of this section. These movements will strengthen your shoulders (deltoid muscles), your arms, and your wrists.

33: Triceps Extension with Dumbbell

Select the appropriate weight. Put your right knee and right hand on a bench. Brace your body with your left leg, which should be firmly planted on the floor. Hold a dumbbell in your left hand, with your arm bent and your upper arm perpendicular to the floor. Keeping your left elbow close to your body, raise the dumbbell so that your arm is fully extended behind you, parallel to the floor. Lower the weight slowly. This constitutes one repetition. Repeat with the other arm.

This exercise can be incorporated into a regular workout by following the general instructions at the beginning of this section. These movements will strengthen the backs of your arms and your shoulders.

 # 34: *Biceps Curl*
Select the appropriate weight level. Sit on a bench with your legs apart. Lean forward and grasp a dumbbell in your right hand, with the palm turned upward, away from your body. Brace your right elbow against your right leg, just inside the right knee. Keeping your right elbow nestled against your knee, curl the dumbbell up to your chest. Then lower the weight to the starting position. This constitutes one repetition.
This exercise can be incorporated into a regular workout by following the general instructions at the beginning of this section. You'll strengthen your biceps.

This description of flexibility and strength training completes our discussion of the power of lower-intensity exercise. But it is only the first step in becoming a part of the Antioxidant Revolution. Next, we will explore perhaps the most popular, yet perhaps the most poorly understood, subject commonly associated with antioxidants—the issue of vitamin and mineral supplements.

PART THREE

Toward Radical-Free Eating and Breathing

The Antioxidant Cocktail—
Advantages, Side Effects,
and Variations

There was a time when I joined most other mainstream physicians who opposed taking vitamin supplements in *any* amounts, much less in relatively large doses. Along with the majority of the medical establishment, I believed that you could get all the vitamins and minerals you needed through your daily diet.

But my research into free radicals and antioxidants has forced me to change my thinking—as well as my personal health habits. Today, I take a daily "antioxidant cocktail," a minimum multivitamin combination of 400 IU of vitamin E; 1,000 mg of vitamin C; and 25,000 IU of beta carotene. Furthermore, when I am scheduled for a heavy physical workout or for some other situation that I know will produce oxidative stress, I increase those amounts accordingly.

The latest research shows that to build strong protection against free radicals, you need to take in far larger amounts of antioxidants than the official Recommended Daily Allowance (RDA) provides. If you regularly consume five to nine ample servings of fresh fruits or vegetables each day, you may be getting enough vitamin C and beta carotene. But you simply cannot get enough vitamin E from the foods you eat. To get sufficient vitamin E, for instance, you would have to eat more almonds, alfalfa seeds, peanuts, wheat germ, or other vitamin E-rich food each day than your system could easily handle. Specifically, just to get 100 IU of vitamin E daily—which is less than what experts are now recommending as the optimum average daily dosage—you would have to eat two cups of almonds, or nearly seven cups of peanuts, *or* one cup of sunflower seeds! The fat and calorie intake would be enormous. Even to get 1,000 mg of vitamin C, it would be necessary to consume up to fifteen oranges or twenty-five green peppers. Or to get 25,000 to 50,000 international units of beta carotene would require consumption of at least two to three carrots or three cups of butternut squash. The only solution is to assemble and consume your own gender-specific, age-adjusted, activity-keyed antioxidant supplement cocktail.

A Cocktail for the Revolution

The chart later in this chapter of antioxidant recommendations reflects the latest information available for the most beneficial amounts of vitamin C, vitamin E, beta carotene, and the mineral selenium. You will note that the amounts vary according to several categories—including age, gender, exercise level, and body weight. All of theses categories and the amounts of antioxidants assigned to them have been suggested by scientific research.

Before you rush out to buy any of these supplements, however, read the following guidelines about antioxidant supplement usage.

Guideline #1: Be Aware of Possible Side Effects

Most people seem to experience no side effects from taking vitamins E, C, and beta carotene in the amounts recommended in this book. But for some people, taking one or more of these antioxidants can pose possible discomfort or even danger. (For additional information on side effects, see also pages 168–169 and 215.)

Vitamin E supplementation is *not* recommended for those patients on anticoagulant therapy. In other words, if you are taking prescribed medications such as Coumadin or aspirin, which work against the clotting mechanism in your blood—as may be the case with those having some type of heart problem—you should avoid supplemental vitamin E completely, or at least consult with your physician before taking it. Vitamin E, which is itself an anticoagulant, may cause too much of an anti-clotting effect in your blood.

In addition, studies conflict on whether or not large doses of vitamin E are associated with an increase in plasma lipids. To be completely safe, it is advisable for those taking vitamin E to go in for regular (annual or semiannual) blood tests to check cholesterol and other lipids.

Caution: Always inform your physician about vitamins or other supplements you are taking. This information is essential to help you guard against dangerous drug interactions. So whenever your doctor asks you what *drugs or medications* you are taking, always mention your vitamin and mineral supplementations too.

Vitamin C in relatively large doses—usually identified as 4,000 mg or more daily—may cause loose bowels or diarrhea in some people. If you notice this side effect, you should cut back on the dosage until the intestinal problems disappear. Those with a history of kidney stones may also have to be careful about taking excess vitamin C. The reason: Vitamin C is water-soluble, so it is excreted through the kidneys and may increase the risk of kidney stones.

Also, be careful about the potential harmful effects of *chewable* vitamin C. This product can make your mouth acidic enough to start dissolving tooth enamel! The researchers who have studied this problem recommend that people buy vitamin C tablets that can be swallowed. (See the *American Journal of Dentistry*, 1992, vol. 5, p. 269.)

Toxic effects of beta carotene in normal, healthy people have not been reported unless the provitamin is taken with relatively large amounts of alcohol or by heavy smokers. This conclusion is based on a 1993 study at New York's Bronx Veterans Affairs Medical Center involving "alcoholic" baboons. The animals were given excessive amounts of alcohol plus 30 milligrams a day of beta carotene (the equivalent of 50,000 IU per day). The baboons taking alcohol and beta carotene suffered more severe liver damage than did another group taking only alcohol.

How much alcohol can you take safely with beta carotene? There are no definitive studies that answer this question right now, but here is my advice, based on the best information available.

If you take beta carotene, I recommend that you drink no more than one ounce per day of pure alcohol—which is the equivalent of about two average-sized, four-ounce glasses of wine, two beers, or one mixed drink. *Under no circumstances* should you take beta carotene if you drink heavily (in the range of four to six ounces of pure alcohol per day).

Also, even if you drink sparingly or moderately, take beta carotene at least four hours after or before you drink alcohol. For example, it is better to take beta carotene with your breakfast, and if you drink, do your drinking at night. Also, beta carotene may be harmful if you are smoking more than one pack of cigarettes per day.

Guideline # 2: Choose Natural Vitamin E

A number of studies, referred to in the January-February 1993 *Nutrition Action Health Letter,* indicate that natural vitamin E is used better by our bodies than synthetic vitamin E. For example, researchers at the National Research Council of Canada found in animal studies that more vitamin E was present in the various organs, tissues, and fluids of those subjects taking natural than was present in those taking synthetic supplements of the vitamin. Also, the Canadians confirmed that more natural vitamin E was likely to be in human blood and plasma than was the synthetic version.

Natural vitamin E is extracted from vegetable oils, such as soybeans, whereas the synthetic supplements come from petroleum or turpentine. How can you tell the difference in your supplements? Just check the labels when you buy vitamin E—but do not rely on a statement that the product contains "natural vitamin E." Such wording may allow supplements that contain only 10 percent of the natural vitamin and 90 percent of the synthetic. Instead, look for the technical name of the vitamin. True natural vitamin E is called "d-alpha tocopherol" or "d-alpha tocopheryl." In contrast, synthetic vitamin E will have an "l" after the "d," as in "dl-alpha tocopherol," or "dl-alpha tocopheryl."

You may also find the words "acetate" or "succinate" after the vitamin E designation. These terms designate organic molecules in our bodies, which are found in energy-producing reactions in every cell. Some regard these forms of vitamin E to be preferable because they tend to be more biologically active.

Even though many experts believe that natural vitamin E may be better absorbed and utilized by the body, one recent study has shown that synthetic E was equally effective in protecting the LDL ("bad") cholesterol from oxidation in sixteen men and women who took 1,600 IU of vitamin E every day for eight weeks. (See *Arteriosclerosis and Thrombosis*, 1993, Vol. 13, p. 601.) But still, the weight of current evidence causes me to continue to recommend that you use the natural form of vitamin E.

Guideline #3: Use Beta Carotene, Not Vitamin A

You should only buy beta carotene, which is a *precursor* of vitamin A—not vitamin A itself. Also called a provitamin, beta carotene is found naturally in yellow and dark green vegetables, such as carrots, pumpkins, sweet potatoes, yellow corn, collards, kale, spinach, and turnip greens. Beta carotene may be converted through a number of biologic processes into vitamin A.

Fully formed vitamin A is found naturally in animal foods, such as liver, butter, and eggs. When taken in doses larger than 5,000 IU, however, vitamin A can be toxic in some people. Among other things, it may cause blurred vision, loss of hair, enlargement of the liver and spleen, or birth defects.

So when you buy beta carotene, do not get a combination tablet that also includes fully formed vitamin A. The only combination "cocktail" you should consider is one that includes the three main antioxidants—vitamin E, vitamin C, and beta carotene—and possibly selenium in amounts no larger than 50 to 100 micrograms.

Guideline #4: Men Need More Vitamin C Than Women

Several studies have reported that men need more vitamin C than women, perhaps because males are larger on average than females and thus require more nutrients overall. Two separate studies published in issues of the *Journal of Clinical Nutrition* in 1987 and 1988 found that men needed three times as much dietary vitamin C as women to maintain high plasma levels of the vitamin.

Also, studies on antioxidants and cataracts—reported from the Laboratory for Nutrition and Vision Research, USDA Human Nutrition Research Center on Aging at Tufts University—suggest that to delay the onset of cataracts, men should consider taking in more than 500 mg of vitamin C per day, and women should take in more than 200 mg per day.

In line with those and related findings, I recommend that all heavy people—both men and women who weigh more than two hundred pounds—increase their intake of all the antioxidant supplements.

Guideline #5: If You Exercise Moderately to Heavily, You Should Increase Your Intake of All Recommended Antioxidants

Because exercise, especially relatively heavy exercise, produces extra free radicals, the body needs more antioxidants to defend itself against damage. Dr. Lester Packer, for instance—the premier free radical and antioxidant researcher from the University of California, Berkeley—concluded in a 1991 report in the *American Journal of Clinical Nutrition:* "Because the primary antioxidant vitamin E is consumed by body tissues during increased physical exercise, results suggest that there is an increased vitamin E requirement during endurance training."

Packer cites a study of humans involved in strenuous exercise who significantly reduced their pentane production (which is evidence of free radical activity in expired breath) by taking 1,200 IU of vitamin E for two weeks. In line with those findings, I recommend that strenuous exercisers, as well as people weighing more than two hundred pounds, take 1,200 IU of vitamin E daily.

Guideline #6: Take Your Vitamins with Meals and, if Possible, at Several Times During the Day

Dr. G. E. Desauiniers of the Shute Medical Clinic in Canada has reported that vitamin E is absorbed better when taken with meals rather than on an empty stomach. Although findings in this area are sparse, I suggest that you take all your antioxidant supplements with meals so that they can be absorbed along with your food.

In addition, there is some indication that large amounts of vitamin C taken in one sitting may be washed out in the urine or feces. So if possible, buy your vitamins in smaller capsules or tablets so that you can take your full dosages in two to three increments throughout the day. That is especially advisable for vitamin C in doses larger than 500 mg.

Guideline #7: Pay Attention to the Measurements of Amounts on the Vitamins You Purchase

It is important not to mix up measurement on the vitamins you buy; otherwise you may end up taking too much or too little. Here is a summary of what you should look for:

Vitamin C is usually sold in milligrams, but if you buy this vitamin in very large amounts, you may also find references to *grams*. Just remember that 1,000 milligrams = 1 gram.

Vitamin E is usually sold in "international units," but if you happen to buy a bottle that uses milligrams instead, remember that one milligram of vitamin E is approximately equal to one international unit (IU) of the vitamin.

Beta carotene is also usually sold in international units, or "IU." But sometimes, you will find a reference to milligrams of beta carotene, and that can create more of a problem in translation with vitamin E.

One IU of beta carotene is equal to 0.6 micrograms of the nutrient, or 0.0006 milligrams. (There are one thousand micrograms to the milligram, and one million micrograms to the gram.) So to convert IU of beta carotene to milligrams, you have to multiply the IU by 0.0006. Or to convert milligrams to IU of beta carotene, divide the milligrams by 0.0006.

Suppose a bottle contains 25,000 IU of beta carotene. You can convert that to milligrams by multiplying by 0.0006, which would give you a result of 15 milligrams. If you read about a study where participants have taken daily supplements of 30 milligrams of beta carotene, you can divide by 0.0006 to get the 50,000 IU equivalent.

Guideline #8: Check the Expiration Date Before You Buy

Rules governing expiration dates for vitamin and mineral supplements are currently much less stringent than those for other foods and drugs. The date of expiration on a vitamin bottle is more or less up to the manufacturer. Still, it is worth looking at the dates so that you can avoid buying any supplement that has an expiration point only six to nine months away. If possible, try to find supplements that will not expire for a year or more. That way, you can be fairly certain that they are reasonably fresh.

Guideline #9: Focus on Getting the Minimum Antioxidants Recommended for You

The main problems associated with taking vitamins C, E, and beta carotene are related to the side effects mentioned in Guideline #1. If you are following the recommendations and you do not notice any of these side effects, it is highly unlikely you are taking too large a dose of the antioxidants, either in supplement or dietary form. On the other hand, if you are taking too little, you

Daily Vitamin and Mineral Recommendations

Age:	5-12	13-21	22-50	50 plus	Heavy exercisers, or weight over 200 lbs.
Vitamin C					
Women:	500 mg	500 mg	1,000 mg	1,000 mg	2,000 mg
Men:	500 mg	1,000 mg	1,500 mg	2,000 mg	3,000 mg
Vitamin E					
Women:	200 IU	400 IU	400 IU	600 IU	1,200 IU
Men:	200 IU	400 IU	400 IU	600 IU	1,200 IU
Beta carotene					
Wom:	10,000 IU	25,000 IU	25,000 IU	50,000 IU	50,000 IU
Men:	10,000 IU	25,000 IU	25,000 IU	50,000 IU	50,000 IU
Selenium (optional)					
Wom:	50 mcg	50 mcg	50 mcg	50 mcg	100 mcg
Men:	50 mcg	50 mcg	50 mcg	50 mcg	100 mcg

may not be getting the full benefits of the "antioxidant cocktail."

So unless you notice one of the side effects—or a qualified physician cautions you to stop taking the supplements because of some health condition or drug you are taking (such as blood thinner)—focus mainly on meeting the minimum recommendations in the chart. The primary reason to avoid taking in more than these amounts through supplements is that current scientific studies show no benefits for higher dosages. With those guidelines in mind, refer to the accompanying chart to find the antioxidant recommendations that are right for you.

Although the guidelines preceding the antioxidant recommendations chart may have answered many of your questions, several others have been raised with regard to the supplements. In particular, I hear many queries regarding how antioxidant supplements may affect cholesterol and blood pressure levels; the risk to health posed by smoking; and the operation of the immune system.

Can Antioxidants Affect
My Cholesterol Levels?

The most important thing to remember about taking antioxidants is that *they work to prevent heart disease regardless of any changes that may occur in your cholesterol levels.* At this point, the primary hero in quenching the free radicals that oxidize the "bad" LDL cholesterol and promote the formation of plaque, which clogs the arteries, is vitamin E. Increasingly, evidence also suggests that vitamin C helps promote the protective work of vitamin E. As I have emphasized throughout this book, that statement represents the latest thinking among scientists.

At the same time, I am *not* saying, nor are most of the other experts, that our new understanding about the role of antioxidants in preventing heart disease means that your cholesterol level is no longer important. On the contrary, an antioxidant program must be combined with a low-fat diet and other cholesterol-lowering strategies.

In fact, antioxidants seem to provide a double benefit for some people as far as cholesterol levels are concerned. Not only are these individuals protected against free radicals, but they also find that their total cholesterol is lowered and their "good" HDL cholesterol is raised.

Before we go any further with this topic, however, let me refresh your memory with a brief reminder about cholesterol terminology, and the lipid (blood fat) scenario occurring in your body. (For more information, see my book *Controlling Cholesterol*, Bantam, 1989.)

Your *total cholesterol* is a blood fat reading, expressed in terms of "milligrams per deciliter," or mg/dl. In general it is best for your total cholesterol to be below 200 mg/dl—though there are more specific optimum values for men and women of different ages, which are listed in charts in *Controlling Cholesterol*. Studies have established that as your total cholesterol goes up 1 percent above 200, your risk of having heart or cardiovascular disease increases by 2 percent.

Your total cholesterol is usually composed mainly of "low-density lipoproteins," or *LDL cholesterol*. This is often referred to as "bad" cholesterol because these particles are the ones that are oxidized by free radicals and combined with macrophages (a form of white blood cell) to form plaque and clogging of the arteries. In general, your LDL cholesterol should be below 130 mg/dl, though again, the number varies somewhat according to age and gender.

Another important component of your total cholesterol is "high-density lipoproteins," or *HDL cholesterol*. Often called "good" cholesterol, high levels of these particles have been associated with a lower risk of cardiovascular disease and heart attacks. HDL cholesterol should be 45 mg/dl or higher for men, and 55 mg/dl or higher for women. Ideally, men and women need higher levels for optimum protection.

Finally, it is important to know the *ratio* of your total cholesterol to your HDL cholesterol. An excellent ratio should be no higher than 4.0 for men and 3.2 for women—and even lower for males and females thirty-nine years of age or younger.

Here is how the ratio is calculated: If your total cholesterol is 180 and your HDL cholesterol is 60, you would set up the ratio as a fraction, 180/60; then divide by 60 to get a ratio of 3.0.

Even though the total cholesterol/HDL ratio continues to be of great importance, more attention has focused recently on the absolute level of the HDL cholesterol. Some investigators believe that an HDL cholesterol of less than 35 mg/dl is probably the best single predictor of future coronary events.

Further studies being performed by Dr. Scott Grundy at the University of Texas Southwestern Medical School in Dallas identify what has been referred to as a "deadly quartet," which is a combination of high blood glucose (sugar), high triglycerides, elevated blood pressure, and low HDL cholesterol in the same patient. Two additional factors contributing to this problem are being overweight and being inactive, but genetic factors also seem to be involved.

In the past, we thought that the only way for most people to lower their total cholesterol and "bad" LDL cholesterol was to consume a low-fat diet, especially one low in saturated fats and foods high in cholesterol. Also, we taught that "good" HDL cholesterol might be raised by endurance training such as jogging several miles per week. Only recently have we begun to look more seriously at a series of studies that indicate antioxidants may have an effect on cholesterol levels in some people.

For example, David Trout, a researcher with the U.S. Department of Agriculture, summed up a number of pertinent cholesterol findings in 1991 in the *American Journal of Clinical Nutrition*. He noted that for more than a decade, it has been known that daily intake of 500 mg to 1,000 mg of vitamin C lowers total cholesterol in a majority of people who have either a low intake of vitamin C in their diets or elevated levels of total cholesterol. In the three studies he cites, the decrease in total cholesterol in people taking vitamin C exceeded 10 percent.

Another study by C. R. Spittle, published in the British medical journal *Lancet* in 1971, found that giving 1,000 milligrams a day of vitamin C to a group with normal cholesterol levels reduced total cholesterol an average of 8 percent. But other similar studies have not reported a similar effect.

As for the effect of vitamin C on "good" HDL cholesterol, three surveys cited by Dr. Trout reported a positive correlation between the amount of ascorbic acid (vitamin C) in the blood and the levels of HDL cholesterol. Two of the studies involved older men, and one involved older women. On the other hand, studies involving younger men in their thirties have not demonstrated this correlation.

To sum up, then, we have some preliminary scientific investigations that suggest vitamin C may lower total cholesterol and raise "good" HDL cholesterol, especially in men and women in their fifties or older. In fact, I have encountered some clinical evidence—involving the family of one of my patients who was among the first to apply the principles of the Antioxidant Revolution—which supports the direction in which these studies are pointing.

A Family Response:
A Case Study of the Power
of Antioxidants Over Cholesterol

My patient, whom I will call Jim, is a fifty-two-year-old New Yorker who had been wrestling with problems of very low "good" HDL cholesterol for many years. His forty-seven-year-old brother, Kevin, from Montana, was also bothered by marginally low HDL cholesterol and high levels of total cholesterol. Their seventy-seven-year-old mother, Grace, from Georgia, was challenged both by high total cholesterol and low HDL cholesterol.

The Georgia Mother

Grace faced a significant total cholesterol problem. A blood test conducted in October of 1992 showed that she had a cholesterol of 259 mg/dl—a result that placed her in a high risk category for developing heart disease. At her age, to be adequately protected from developing cardiovascular diseases, her total cholesterol should have been below 227.

Her blood lipids also produced other concerns. Her "good" HDL cholesterol was only 45—again, a result that placed her at high risk for coronary artery problems. To have optimum protection, her HDLs should have been above 74.

And there was still another problem. Her "bad" cholesterol, or low-density lipoprotein (LDL), was 163, another high reading that placed her in the moderate risk category for heart disease. Her LDLs should have been 149 or lower to afford her excellent protection.

Finally, the ratio of Grace's total cholesterol to her HDL cholesterol was 5.76, a result that put her in the very high risk category for developing heart disease. The ratio should have been 3.2 or lower to provide her with excellent protection.

Actually, Grace had been on a regular, vigorous walking program for more than a year. She walked steadily at a brisk pace for more than an hour per day, five to six days a week. Such endurance exercise often raises HDL levels considerably. But in her case—even though she felt much better and found she had considerably more stamina—the workout did not improve her HDLs.

Her physician was sufficiently concerned about the imbalance in her cholesterol to prescribe a cholesterol-lowering medication, Zocor. The drug succeeded in lowering her total cholesterol dramatically to 194 and her "bad" LDL cholesterol to 90. Also, her "good" HDL cholesterol rose to 62. With those results—and a new 3.1 ratio—she now enjoyed excellent protection from heart disease.

Grace's cholesterol profile had improved so impressively that her doctor allowed her to cut her Zocor medication in half. But then her total cholesterol began to drift above 250 again, and her HDLs went down, which caused a rise in her ratio. In other words, her risk of developing heart disease increased again.

It was at that point that her son Jim, the New Yorker, told her about the potential benefits of antioxidants—specifically, vitamins E, C, and beta carotene. Within a week, she began to take daily doses of 1,000 IU of E; 500 milligrams of C; and 25,000 IU of beta carotene. (Those amounts are somewhat lower than what I recommend for a woman of her age and activity level. Specifically, I tell heavy female exercisers of all ages to take 1,200 IU of E; 2,000 mg of C; and 50,000 IU of beta carotene.)

The results with even the smaller amounts were highly encouraging. Grace's latest cholesterol profile, which was taken in September 1993, about five months after she began the antioxidants, looked like this:

Total cholesterol: 228
HDL (good) cholesterol: 102
LDL (bad) cholesterol: 77
Total cholesterol/HDL ratio: 2.24

The results were remarkable. Her HDL cholesterol increased by more than 65 percent over the level at which it had been before she began the antioxidant program. Also, her ratio decreased to an even lower and safer level. Even while she was only on the cholesterol-lowering medication, her HDL cholesterol had been lower and her ratio higher than when she was on the Zocor plus the antioxidant supplements.

You will note that the total cholesterol may still seem slightly high, but a large portion of that total consists of the "good" HDL cholesterol and very little of the "bad" LDL cholesterol. I never worry about the total cholesterol if a person has a lipid profile like Grace's.

Why did those beneficial changes occur? It is possible that her very high HDL cholesterol, her very low LDL, and the excellent ratio were affected by the antioxidant supplements. It is also possible that the supplements worked in tandem with the Zocor prescription to improve the cholesterol balance.

Yet I want to emphasize the *possible* effect of antioxidants favorably affecting the cholesterol levels. There is still no definitive scientific evidence that they will correct an abnormal lipid profile.

The New York Son

Blood tests on fifty-two-year-old Jim, who has been my patient for twelve years, consistently showed low levels of HDL cholesterol, usually in the mid-thirties range. For example, in February 1992, his HDL value was 36—a level that placed him in the high coronary risk category for his age.

Low doses of the over-the-counter vitamin niacin (1,000-1,500 mg per day), which is a powerful cholesterol-lowering drug, helped him keep his cholesterol below 200 on most visits to my office. But his unusually low HDL was unaffected by the medication and, as a result, his ratio was quite high, putting him into the high to very high risk group.

On that February 1992 exam date, Jim's total cholesterol was a low-risk 196. But the ratio of his total to his HDL cholesterol was 5.4, or well into the moderate risk range.

Jim then embarked on my Antioxidant Revolution program. He had been exercising regularly for nearly twenty years, alternating between slow jogging and racquet sports, three to four times a week. In addition, he pursued a regular strength training program, consisting of calisthenics and some weight work, three to four times a week. To his exercise, he added daily antioxidant vitamin supplements in these amounts in early August of 1992:

25,000 IU of beta carotene
400 IU of vitamin E
1,000 mg of vitamin C

I should note that he had taken two of these vitamins with some regularity before he increased them to the above amounts. Specifically, he usually took 400 IU of vitamin E and 500 mg of C. I had thought that those relatively limited doses probably would not work with him, and in fact, they seemed to have no effect on his cholesterol profile.

But things changed after Jim increased his antioxidant dosage. When he went in for a blood test about five weeks after the commencement of his new antioxidant program, his cholesterol profile looked like this:

Total cholesterol:	205
HDL cholesterol:	45
LDL cholesterol:	140
Total cholesterol/HDL ratio:	4.6

In spite of the fact that his total cholesterol was slightly higher than it had been at his previous blood test, Jim's ratio was now much lower, close to the low-risk threshold for his age, which is a ratio of 4.2. The key to that transformation was the dramatic rise in that "good" HDL cholesterol. He had experienced an amazing 25 percent increase in his HDLs!

The improvement continued apace through Jim's next physical exam, which was conducted in April 1993. The blood tests taken at that time revealed the following values:

Total cholesterol: 194
HDL cholesterol: 47
LDL cholesterol: 121
Total cholesterol/HDL ratio: 4.1

This time, Jim's total cholesterol had dropped, as had his "bad" LDL cholesterol. Since his diet had not changed during that period, it was quite likely that the antioxidant program was triggering these reductions.

At the same time, his HDL readings had gone up further, to a point that was by far the highest we had ever recorded on this patient. At 47 mg/dl, the HDLs had moved into a low moderate risk range and were closing in on the low risk level that begins at 52 mg/dl. There was now a 30.5 percent increase in HDLs over the blood test conducted just before Jim had started on the full antioxidant supplement dosage. Even more important, Jim's total cholesterol/HDL ratio was now in the low coronary risk category.

What caused the great improvement? It appears that the antioxidant program played a major role in raising the HDL levels, improving the ratio—and perhaps also lowering the total cholesterol and "bad" LDL cholesterol. As with his mother, Jim's medicine (niacin) may have interacted with the vitamin supplements to help produce the change in the cholesterol levels. Or the antioxidants may have done the job by themselves.

To see what might happen to a person who takes antioxidants without a cholesterol-lowering drug, I asked for the blood profile on Jim's brother, Kevin, who lives in Montana.

The Montana Son

Kevin, who was forty-seven, was not taking any medications when he went in for three blood tests through a four-and-a-half year period. The first of those tests, in June 1989, showed these results:

Total cholesterol: 254
HDL cholesterol: 51
LDL cholesterol: 168
Total cholesterol/HDL ratio: 4.98

As you can see, the total cholesterol was high and placed him in the high-risk category for coronary artery disease. In fact, Kevin just missed the very high risk level for his age, which begins at 257. His bad LDL cholesterol was also in the high-risk area.

On the other hand, unlike his brother and mother, Kevin's good HDL cholesterol was closer to being normal—though it still put him in the moderate-risk category. Apparently he does not suffer from the genetic problem that keeps the HDLs of his other relatives at abnormally low levels. Because his HDLs were higher, Jim's ratio was better than his brother's earlier exams, though he still ended up in a moderate coronary risk category with this particular ratio.

At this time, Kevin was doing only minimal exercise and paying little attention to his diet. Also, he was taking no antioxidant supplements. But soon after this test, he became more active by jogging two miles, two to three times a week, and hiking in nearby mountains at least once a week. When he had his blood tested in October 1989, there were some slight changes in his blood lipids:

Total cholesterol: 237
HDL cholesterol: 52
LDL cholesterol: 161
Total cholesterol/HDL ratio: 4.6

The one-point increase in his "good" HDL cholesterol may have been due to the increased exercise or perhaps to a laboratory variation. In any event, at 52 mg/dl, he was now in the low coronary risk, "excellent protection" category for the HDL reading. Kevin's apparent response to endurance exercise is common in many people, though neither his mother nor his brother, both of whom worked out regularly, showed that capacity.

Kevin's weight and diet remained constant during that six-month period, but since he was exercising considerably more than before, we can assume that his muscle mass increased and his body fat decreased. Such a shift would explain the lower total cholesterol and lower LDL cholesterol, both of which tend to rise with increased body fat. Because the total cholesterol went down and the HDLs went up, Kevin's ratio improved to some extent.

By 1993, Kevin was considerably more sedentary than he had been in October 1989. He had put on two pounds, was eating extra desserts (including frequent pound cakes and pumpkin pies), and reported that he was doing "no exercise" and working long hours. But he did go on a daily antioxidant supplement program in April of 1993. Specifically, he began to take 1,000 milligrams of vitamin C (which he reduced to 500 milligrams in July of 1993). He also took daily doses of 25,000 IU of beta carotene, and 400 IU of vitamin E.

When Kevin went in for a blood test in October 1993, these were the results:

Total cholesterol:	248
HDL cholesterol:	52
LDL cholesterol:	166
Total cholesterol/HDL ratio:	4.77

There was little change in his blood profile since October 1989, even though he had neglected his exercise and dietary habits. His total cholesterol and "bad" LDL cholesterol had gone up slightly, undoubtedly the result of his higher fat diet and increased weight. Yet his HDL cholesterol had stayed the same as when he had been working out regularly.

What are we to make of Kevin's blood test? Favorable changes in his cholesterol (or lipoprotein) blood levels are probably not brought about by the use of antioxidant supplements. But certainly, prevention of LDL oxidation in the arterial wall can be expected. Furthermore, I predict that if he had added regular, lower-intensity endurance exercise to his program, his cholesterol blood profile would have looked even better.

A related study performed by Scott Grundy showed that taking three antioxidant vitamins—vitamins E, C, and beta carotene—provides more protection against LDL oxidation than any one vitamin alone. Those findings provide a rationale not only for the beneficial interaction of supplements, but also for using the "triple antioxidant" cocktail to protect against coronary artery disease. But the precise extent to which antioxidant supplements may affect cholesterol levels in different population groups must await further studies.

Can Antioxidant Supplements Lower Blood Pressure?

Beginning in 1978, a series of studies in the United States, Japan, and Finland—involving both men and women of a wide variety of ages—established that higher amounts of vitamin C in the diet result in lower systolic and diastolic blood pressure readings.

A reminder: The systolic pressure, the "upper" or first number in a blood pressure reading, indicates in millimeters of mercury (mm Hg) the force of the blood against the artery wall as the heart beats or pumps. The diastolic pressure is the "lower" or second number, which indicates the force of the blood against the vessel walls when the heart is "resting" between beats. Normal blood pressure is 120/80 mm Hg.

In confirmation of earlier findings, researcher E. T. Koh did a study of twenty-three borderline hypertensive women, which was reported in the *Journal of the Oklahoma Medical Association* in 1984. The systolic pressures of these women ranged from 140 to 160 mm Hg, and their diastolic pressures were between 90–100 mm Hg. Koh gave them an extra 1,000 mg of ascorbic acid (vitamin C) per day for three months and found an average seven-point lowering of systolic pressure and a four-point lowering of diastolic pressure.

Dr. David Trout, of the Carbohydrate Nutrition Laboratory, Beltsville Human Nutrition Research Center, United States Department of Agriculture, conducted an unpublished study involving twelve borderline hypertensive patients, aged thirty-five to seventy-four. They were first given 1,000 mg of vitamin C for a six-week period. After a two-week break they were fed a placebo for six weeks. The high vitamin C regimen resulted in a lowering of systolic blood pressure, though not of diastolic pressure in that group.

What can we conclude from these findings? The use of vitamin C supplements in the amounts recommended in this book— 1,000 mg or more—seems clearly advisable for those with high blood pressure concerns. I can only agree with Dr. Trout, who says:

> Blood pressure appears to be susceptible to effects of extra AA (ascorbic acid, or vitamin C) even when vitamin C status may be considered good or excellent. This finding is new enough and important enough to warrant further verification.

Can Antioxidants in Any Way Protect Cigarette Smokers—and Those Exposed to Sidestream Smoke?

My first and greatest commandment of good health has always been "Don't smoke cigarettes!"—and I still subscribe wholeheartedly to that dictum. But in my research into free radicals, I have been amazed to learn that antioxidants—especially when taken in relatively large doses—may provide smokers with significant protection against lung cancer and free radical damage.

Here is some of the scientific evidence:

Beta carotene. The Multiple Risk Intervention Trial Study, begun between 1973 and 1975 and reported in 1989, measured the relationship between serum levels of beta carotene and the appearance of lung cancer. The researchers found that the beta carotene levels were significantly lower in the smokers who developed lung cancer than in the smokers who did not develop cancer.

According to Anthony Diplock, quoted in 1991 in the *American Journal of Clinical Nutrition,* this study shows that "the results provided further evidence for a protective effect of beta-carotene against lung cancer in smokers."

The much-publicized 1994 study by the National Cancer Institute of male smokers in Finland actually revealed an increase in lung cancer among those placed on large beta carotene dosages. But the participants had each averaged a pack of cigarettes a day for thirty-six years, and they were only on the supplements for a few years.

Vitamin E. In 1991, Lester Packer of the Department of Molecular and Cell Biology, University of California, Berkeley, surveyed the protective role of vitamin E in different biological systems, including the human body.

He reported that in one 1988 study of smokers, the output of pentane in the breath of smokers could be suppressed by taking daily supplements of 800 IU of vitamin E. (You will recall that pentane is a residue that indicates the presence of free radicals.) Another study found that fluids of the lower respiratory tracts of smokers were deficient in vitamin E, but that the deficiency could be improved somewhat with a daily intake of 2,400 IU of vitamin E for three weeks. The researchers suggested that a vitamin E deficiency in young smokers may expose their lungs to increased free radical damage.

In animal studies, Packer said, vitamin E has been shown to protect lungs that are exposed to cigarette smoke or to ozone (the pollutant present in much urban air). Packer concluded that "vitamin E may help protect the lungs from damage associated with exposure to common air pollution."

Also, you'll recall that the National Cancer Institute's study of Finnish male smokers, published in April 1994 in *The New England Journal of Medicine*, showed significant decreases in prostrate cancer, colo-rectal cancer, and heart disease among those taking vitamin E daily.

Vitamin C. Researcher Ronald Anderson—of the Medical Research Council Unit for the Study of Phagocyte Function, Department of Immunology, Institute for Pathology, University of Pretoria, Republic of South Africa—studied the protective role of vitamins C, E, and beta carotene in fighting free radical damage linked to cigarette smoke. He notes in a 1991 report that decreased blood levels of these three antioxidants are associated with cigarette smoking.

Anderson concluded that excess free radicals are associated with inflammatory tissue damage and possibly bronchial carcinoma (cancer) in smokers. But "nutritional antioxidants, vitamin C, vitamin E, and beta-carotene appear to be of some importance in the prevention of pulmonary (lung) damage mediated by these oxidants (free radicals)," he said.

The beneficial possibilities of vitamin C for smokers has been recognized to the point that the offical R.D.A. for vitamin C has been raised from 30 mg. to 60 mg a day for smokers.

Warning: Do not allow these studies to make you complacent if you are a smoker! Although such findings suggest that it is possible to counter the effects of cigarette smoking to some extent by taking relatively large doses of antioxidant supplements, the best approach is still to avoid smoking altogether. Most likely, antioxidants can overcome only part of the damage inflicted by cigarette smoke—not all of it.

Can Antioxidant Supplements Help Fight AIDS?

On the cutting edge of research into antioxidants and free radicals is the issue of AIDS and the virus, H.I.V., which causes the disease. At a conference conducted by the National Institutes of Health in November 1993, scientists reported studies that have linked the AIDS virus to excessive free radical activity. They emphasized that with this disease, free radicals can weaken or destroy the body's immune system.

Papers presented at the meeting included these findings:

- A Harvard study indicated that the use of antioxidants could stop abnormal, destructive communication between the body's cells, which occurs in the presence of the H.I.V. virus.
- A Johns Hopkins School of Hygiene and Public Health study found a deficiency of vitamin A among AIDS patients.
- Scientists at the conference indicated that vitamin therapy, under the direction of a qualified physician, may be appropriate for those with AIDS to combat the increased free radical activity and the depressed immune response associated with the disease. Vitamin supplements that were suggested as part of this therapy included vitamin E, vitamin C, beta carotene, and vitamin A.

Obviously, that report is preliminary, and no one with AIDS or the H.I.V. virus should act on his or her own to try a mega-vitamin treatment of the disease. But a number of physicians are already using a vitamin therapy approach with their AIDS patients. However, the benefit for AIDS patients is quite conjectural at this time.

These and related findings about the significant benefits of antioxidant supplements provide an incontrovertible argument for taking vitamins E, C, and beta carotene in the doses recommended in this book. Many physicians have been taking these supplements themselves for years, even before the weight of scientific authority shifted so heavily in the antioxidant direction. It is virtually inevitable that within the next two to five years, dosages much larger than those included in current RDA amounts will be the standard followed by doctors who keep up with the latest medical developments. Why wait? If you have not done so already, include the new supplement recommendations in your program right now. As the University of California at Berkeley *Wellness Letter* said, "The role these substances play in disease prevention is no longer a matter of dispute."

7

Cooking and Eating
the Antioxidant Way

Although I am certainly not an expert cook, my study of antioxidants in recent years has caused me to look much more carefully at the way my food is selected, stored, and prepared. While a supplementation program is essential for those who want to increase their body's free radical defenses to adequate levels, we have to remember that a "supplement" is an *addition* to a healthy food program. A diet high in antioxidants must remain the cornerstone of the nutrients we take in from the outside. There are also substances in foods other than antioxidants that may provide protection from cancers and heart disease.

As I look over the foods on display in a cafeteria line, I sometimes let my mind drift back to the kitchen and try to visualize how the foods I am about to select have been prepared. A series of questions comes to mind:

- What is the maximum potential antioxidant content of the foods, assuming no loss of nutrients during storage or preparation?
- What kind of commercial packing has been used, and how have the foods been stored before this meal? What is the likely impact of the precooking procedures on the antioxidant content of the foods?
- What techniques of preparation and cooking are being employed? How may those techniques affect the antioxidant content?

- What will be my total intake of antioxidants today, including the supplements I have used, after I finish this meal?

After going through this sequence of queries and making a few estimates, I find that I can usually arrive at a ballpark approximation of the amounts of antioxidants I am taking in each day through my diet and supplement program. By evaluating informally the antioxidant content of each of my meals in advance, I am more likely to be aware of dishes such as carrots, sweet potatoes, or fruits, which can significantly increase my daily intake of natural free radical fighters.

The same can be true for you—if you can learn to approach each meal in a systematic way that promotes the Antioxidant Revolution in your life. To help you evaluate your own diet, let me suggest that, whenever feasible, you take the steps outlined below before a meal.

This strategy can be used at home or in a restaurant or cafeteria—though if you are eating out, you will have to ask your waiter, the manager, or the chef some pointed, specific questions about meal preparation. After eating in a restaurant or cafeteria several times, you may find that the waiters and the chef are quite willing to disclose information about where they buy their food and the basic preparation techniques used by the establishment. Usually, a good waiter is eager to share his or her knowledge about what is fresh that day and what has been frozen or stored for a time.

Step One:
Estimate the Maximum Potential Antioxidant Content of the Food

With this first step, you don't worry about how much antioxidant content may have been lost during shipping, storage, or preparation. Simply look at the nature of the food and the quantity, and estimate what the maximum amount of vitamin C, vitamin

E, or beta carotene in it would be if none of the nutrients had been lost.

Although many, many foods contain some antioxidants, only a few have sufficiently large quantities to be considered candidates for your antioxidant program. Here are the foods that you should try to make staples of your diet. They are grouped under each of the main antioxidant vitamins.

Foods with High Beta Carotene Content

Common foods that tend to be high in beta carotene, the precursor or "provitamin" of vitamin A, include sweet potatoes, carrots, cantaloupes, pumpkin, butternut and winter squash, spinach, broccoli, mango, and papaya.

As you know from chapter 6, the minimum amount of beta carotene every adult should be taking in each day is 25,000 IU, with certain groups, such as heavy exercisers, getting 50,000 IU. What would you have to eat to get beta carotene in those amounts from your food, without taking supplements? Here are some possibilities:

- You could get 50,000 IU of beta carotene from one cup of cooked sweet potatoes, three medium carrots, or one cup of cooked pumpkin. (This is the equivalent of about 30 mg of beta carotene.)
- You could get approximately 25,000 IU of beta carotene (15 mg) from one-half of a medium, cooked sweet potato; one-half a cup of cooked pumpkin; one and a half medium-sized carrots; one and a half cups of cooked spinach; or two medium-sized mangoes.
- If you want to mix up your vegetable diet, as you probably do, you could also achieve about 25,000 IU of beta carotene with these combinations:
  1. one-half cup of boiled spinach, one cup of cantaloupe slices, and one-half a raw carrot

2. one-half cup of boiled broccoli, three medium apricots, and one-half cup of boiled carrot slices

3. one medium papaya, one-half cup of frozen spinach, and one-half baked sweet potato.

Now, consult the accompanying extensive list of foods containing relatively high amounts of beta carotene. The values are expressed in International Units (IU). Abbreviations used in these tables have the following meanings:

oz. = ounces
msh. = mashed
c. = cup
cnd. = canned
frzn. = frozen

Note: When foods are listed as "boiled" or "raw," several assumptions are made: Minimal amounts of water are used, the water is brought to a boil before the foods are put into the water, and the cooking time is short, so that vegetables remain crisp. The water used for boiling is saved and served with the food (nutrients that may escape during the boiling process still tend to stay in the water). Raw foods are served immediately after purchase and not allowed to wilt, to be exposed for more than a day to air or sunlight, or to dry out. In measuring foods, larger amounts of cooked foods usually fit into a cup than raw foods.

Finally, remember that beta carotene is found in plant foods, whereas fully formed vitamin A is found in animal foods, such as liver. For the Antioxidant Revolution, you need beta carotene, not vitamin A. As a result you will only find fruits and vegetables in this list.

Foods with High Vitamin C Content

If you want to try something exotic and get a very high dose of vitamin C for a relatively small intake of food, try the fruit of the acerola shrub. This West Indian plant—which certainly isn't a

List of International Units (IU)
of Beta Carotene in Selected Foods

Foods	Beta carotene International Units (IU)
Apricot, 1/2 c. frzn. sweetened	2,033
Apricot, 3 medium	2,769
Apricot, dried 10 halves	2,534
Asparagus, 6 spears boiled	746
Broccoli, 1/2 c. raw	678
Broccoli, 1/2 c. boiled	1,099
Broccoli, 1/2 c. frzn., chopped, boiled	1,741
Brussels sprouts, 1/2 c. boiled	561
Cantaloupe, 1 c.	5,158
Carrot, 1 raw	20,253
Carrot, boiled 1/2 c. slices	19,152
Dandelion greens, 1/2 c. raw	3,920
Green peas, 1/2 c. frzn. boiled	534
Kiwi, 1 medium	133
Mango, 1 medium	8,060
Mustard greens, 1/2 c. boiled	2,122
Mustard greens, 1/2 c. frzn. boiled	3,352
Papaya, 1 medium	6,122
Parsley, 1/2 c. raw	1,560
Seaweed, laver (nori) 3.5 oz. raw	5,202
Spinach, 1/2 c. raw	1,880
Spinach, 1/2 c. boiled	7,371
Spinach, 1/2 c. cnd.	9,391
Spinach, 1/2 c. frzn.	7,395
Squash, butternut 1/2 c. boiled	7,141
Squash, winter 1/2 c. baked	3,628
Sweet potato, baked	24,877
Sweet potato, 1/2 c. msh.	27,968
Sweet potato, 1 c. cnd.	15,965
Tangerine, 1 raw	773
Tomato, 1 raw	1,394
Turnip greens, 1/2 c. raw	2,128
Turnip greens, 1/2 c. boiled	3,959
Watermelon, 1 c. raw	585

household name in the United States—has mildly acidic, cherry-like fruits that are packed with the vitamin. One cup of raw acerola fruit contains more than 1,600 mg of vitamin C, and eight fluid ounces of fresh acerola juice has an astounding 3,800 mg! (Remember, my basic recommendation for daily intake of vitamin C is 1,000 mg. The dosage goes up to as much as 3,000 mg for very heavy men or men who are involved in regular, heavy exercise.)

If acerola isn't your "cup of juice," or if you decide you do not like it after trying it, there are plenty of alternatives. More common foods that are relatively high in vitamin C content include papaya, black currants, strawberries, oranges and orange juice, cantaloupe, cranberry juice, and grapefruit juice.

Here is a suggestion about how you can quickly and easily put together a 500 mg daily intake of vitamin C in your diet: Drink one and a half cups of orange juice; eat one-half cup of raw broccoli; eat one cup of raw strawberries; and eat one-half cup of raw cauliflower.

Check the a list of foods containing relatively high amounts of vitamin C. The values are expressed in milligrams (mg). Remember that vitamin C is found mainly in fruits and vegetables. As a result, you will only find fruits and vegetables in this list. Abbreviations used in these tables have the following meanings:

oz. = ounces
msh. = mashed
c. = cup
cnd. = canned
frzn. = frozen
fl. = fluid
ckd. = cooked

Foods with Vitamin E Content

Unlike beta carotene and vitamin C, vitamin E is not found in high quantities in foods that you can eat easily in one day's bal-

List of Milligrams (mg) of Vitamin C in Selected Foods

Food	Vitamin C mg
Acerola juice, fresh 8 fl. oz.	3,872
Acerola, raw 1 c.	1,644
Asparagus, boiled 1/2 c. raw (6 spears)	18
Broccoli, 1/2 c. raw	41
Broccoli, 1/2 c. boiled	49
Broccoli, 1/2 c. frzn. boiled	37
Brussels sprouts, 1/2 c. boiled (4 sprouts)	48
Cantaloupe, raw 1 c. pieces	68
Cranberry juice cocktail, bottled 8 fl. oz.	108
Cauliflower, 1/2 c. raw	36
Cauliflower, 1/2 c. boiled	34
Grapefruit, raw 2 pink	47
Grapefruit juice, fresh 8 fl. oz.	94
Grapefruit juice, cnd. 8 fl. oz.	72
Guava, 1 medium raw	165
Honeydew melon, 1/4 c. chopped	23
Kiwi, 1 medium raw	75
Lemon juice, fresh 8 fl. oz.	112
Mango, 1 medium raw	57
Orange, navel 1 raw	80
Orange juice, fresh 8 fl. oz.	124
Orange juice, cnd. 8 fl. oz.	86
Orange juice, frzn. concentrated 8 fl. oz.	97
Papaya, 1 medium raw	188
Pepper, 1/2 c. green chopped	45
Pepper, 1/2 c. red chopped	95
Strawberries, 1 c. raw	85
Strawberries, frzn. sweetened 1 c.	106
Tomato juice, 6 fl. oz.	33
V8-Juice, 6 fl. oz.	37

anced diet. For example, here are some comparisons of foods with the highest vitamin E content. (The vitamin E values may be expressed either in milligrams or international units, depending on whether the item is in oil or solid food form. But remember that one milligram of this vitamin is about the same as one international unit, or IU.)

One tablespoon of wheat germ oil = 20.3 mg of vitamin E
One ounce or 28 grams of almonds = 10.10 IU of vitamin E
One-third cup of toasted wheat germ = 6 IU of vitamin E
One tablespoon of Hellmann's Mayonnaise = 11 mg of vitamin E
One sweet potato = 5.93 mg of vitamin E

You can see that it would take impossible amounts of these foods each day even to reach the minimum 100 IU of vitamin E that you need to fight free radicals. Furthermore, these nuts and oils have extremely high levels of fats, which are unacceptable on the low-fat diet that the American Heart Association and other recognized medical organizations have established. No one should be taking in more than 30 percent of his or her calories each day in the form of fat, and an intake of 20 percent or below is preferred.

The only answer to this dilemma is to choose vitamin E foods whenever possible if you can do so without violating a healthy low-fat regimen, and then to make up the difference with a natural vitamin E supplement of the type described in the previous chapter.

For your information, I've included a list of foods with the most vitamin E. Vitamin E in solid processed foods is usually expressed in international units, whereas vitamin E in oils is usually expressed in milligrams.

Some abbreviations used in this table include the following:

oz. = ounce
g = grams
c. = cup

List of Vitamin E in Selected Foods
in Terms of International Units (IU) or Milligrams (MG)

Foods	Vitamin E International Units (IU)
Almonds, blanched 1 oz.	8.75
Almonds, whole 1 oz. 28 g	10.10
Breakfast bar	
Carnation choc. chip 1 bar	7.50
Carnation choc. crunch	7.50
Carnation peanut butter crunch	7.50
Figurine diet bar	
Pillsbury choc. 1 bar	4.18
Pillsbury vanilla	4.20
Hazelnuts, dried, 1 oz.	6.70
Hazelnuts, roasted 1 oz.	4.40
Pudding, Delmark, from mix choc. 1/2 c.	5.40
Slender bars, Carnation choc. 2 bars	7.50
Sunflower seeds, dried, 1 oz.	14.18
Sweet potato, 1 raw	5.93
Wheat germ, 1/3 c. dry mix	6.00
Whole wheat, 1/3 c. dry mix	3.00

	Vitamin E mg.
Wheat germ oil, 1 T	20.30
Almond oil, 1 T	5.30
Corn oil, 1 T	1.90
Corn oil, Mazola 1 T	3.00
Cottonseed oil, 1 T	4.80
Safflower oil, 1 T	4.60
Sunflower oil, 1 T	6.10
Margarine, Mazola, 1 T	8.00
Mayonnaise, Hellmann's, 1 T	11.00

choc. = chocolate
T = tablespoon

Now that you have access to this information about the antioxidant content of various foods, you are ready to take the second step in an Antioxidant Revolution eating program: evaluating the effect of commercial packing, storing, cooking, or other preparation procedures on the antioxidant content of the foods you plan to eat.

Step Two:
Evaluate the Impact of Cooking and Other Food Preparation Procedures on Antioxidant Content

Even if you see that a food is high in an antioxidant, that doesn't mean that when you eat it you will receive the full listed benefits. Considerable nutritional value of foods can be lost during packing, storage, cooking, or other meal preparation procedures. (See appendix 4 for additional information.) In general, antioxidants are susceptible to being reduced by such factors as these:

- Changes in their "pH" value, or acidity versus alkalinity. Such changes may occur when certain additives are included during processing.
- Exposure to oxygen. Packaging may actually help antioxidant retention. For example, vacuum-packed frozen vegetables and fruits have the best retention of vitamin C.
- Exposure to light.
- Exposure to heat or high temperature.
- Exposure in your kitchen to the oxygen in air.

When you freeze fruits or vegetables, the optimum temperature is zero degrees Fahrenheit. Fresh vegetables should be stored in a refrigerator vegetable crisper or sealed in moisture-proof bags to maintain their nutritional quality. Wilted vegetables will lose considerable vitamin C and beta carotene content, in contrast to those that remain fresh.

The best cooking methods to preserve antioxidants are these:

- Microwave
- Steam
- Stir-fry

Now, let's move on to some of the specifics of preparing meals the antioxidant way.

We will begin our discussion of how to minimize the loss of antioxidant content in your foods by focusing on the areas over which you have the most control—kitchen storage and preparation of food. (In appendix 4, I have included some other points that relate to the loss of antioxidants during agricultural and commercial processing of foods.)

Getting Beta Carotene into Your Body

Including vitamin E in your diet or as a supplement can help your absorption of beta carotene.

Note: Taking larger doses of vitamin E—above about 100 IU—may interfere with the conversion of beta carotene into vitamin A in some people. But remember, we are not worried about your supply of vitamin A because this vitamin isn't an effective antioxidant, and plenty of vitamin A is usually available through a normal diet. The important consideration is how much of the precursor of vitamin A, beta carotene, you are getting into your system.

Alcoholic beverages, which contain ethanol, may decrease your body's ability to process beta carotene. Also, the cholesterol-lowering drug Colestipol may have a similar effect.

The absorption of beta carotene in foods varies. For example, you can expect 36 percent in carrots to be used by your body; 46 percent in papaya; and 33 percent from other vegetable sources. The lesson here is that you must figure that only one-third to less than a half of the beta carotene in different foods you eat will actually be able to contribute to your free radical defenses. Hence, there is a need to reinforce your antioxidant intake with supplements.

Cooking affects beta carotene absorption as well, often for the better. With carrots, for instance, only 1 percent of the beta carotene in a raw carrot may be used by your body. On the other hand, mild cooking—such as light steaming—can improve absorption dramatically by increasing the digestibility of the carrot. On the other hand, overcooking green leafy vegetables can decrease the "bioavailability" of their beta carotene—or the amount of the nutrient that can actually be absorbed and used by the body. Steaming or cooking with minimal amounts of water so that vegetables remain crisp is the preferred way to prepare foods high in beta carotene.

Making the Best Use of Vitamin C

In general, vitamin C tends to remain stable in acidic solution, but it decomposes significantly in the presence of alkalies (such as baking soda), oxygen, copper (including cooking utensils), and iron (in food or cooking utensils). Although preservatives sometimes get bad press, sulfite additives can act as scavengers for oxygen; in other words, they fight free radicals. Sulfur dioxide that comes in contact with foods during preparation protects vitamin C as well as beta carotene.

Washing, trimming, peeling, and cutting fruits and vegetables will reduce their vitamin C content, probably because those procedures expose more of the food to oxygen.

Be careful with your broccoli! Many families I know will cut off and cook the florets (the attractive green flower), but will discard the less attractive yet tasty stems. Unfortunately, the stems retain vitamin C much better than the florets because the florets have a greater surface area exposed to air, heat, and water.

Always microwave, steam, or stir-fry foods containing vitamin C. Scientific tests have shown that cabbage and broccoli lose only 10 to 20 percent of their vitamin C during microwaving, as opposed to losing 27 to 62 percent when boiling in large amounts of water. Vitamin C is leached out by water, but it can be recovered to some extent if you use the water in serving your meal. Reheating or holding cooked foods for extended periods causes additional losses of both vitamin C and beta carotene.

Finally, be sensitive to the possible losses of vitamin C at different stages of the harvesting, marketing, and consuming processes. Here are some guidelines:

- "Garden fresh" spinach—or spinach immediately after a farmer pulls it from the ground—may lose 90 percent of its vitamin C within ten days after harvest. Yet "market fresh" spinach—or the spinach sold at the wholesale or retail level—may have been harvested four to thirteen days before it was delivered to the store.
- Broccoli has been shown to lose 17 percent of the vitamin C it had at the wholesale level when it reaches the retailer (such as your supermarket) and 27 percent of the wholesale amount by the time you have stored it for three days.
- The vitamin C values of different foods may also vary markedly. For example, one study showed that New Jersey strawberries contained an average of only 49 mg of vitamin C per 100 grams of weight, as compared with 65 mg in California and Florida berries. Furthermore, although the Florida and California berries retained their vitamin C quite well during storage, the New Jersey berries lost significant amounts of vitamin C over time; they went from 62 mg of vitamin C at the beginning of storage, to 53 grams after four days of storage, to 33 grams after seven days of storage.
- Some vegetables are much better than others in retaining vitamin C. One of the best is the sweet pepper, which can be stored for two to three weeks without significant loss of the vitamin.

Preserving Your Vitamin E

Although you can only expect to get small amounts of vitamin E from your diet, it is still advisable to save as much of it as possible. Here are a few tips:

- Vitamin E becomes unstable and may break up or be destroyed during frying, especially deep-fat frying. So be aware that any frying, including stir-frying, will destroy vitamin E.
- Vitamin E becomes unstable and will be lost at room temperature when it is exposed to oxygen, alkalies (such as baking soda), iron salts (ferric salts), or ultraviolet light.
- Copper or iron utensils may cause loss of vitamin E.
- Adding foods high in vitamin C to cooking or other food preparation can help preserve the vitamin E. For example, you might put wheat germ oil (contains vitamin E) as a dressing on a broccoli salad (contains vitamin C). Or if you plan to stir-fry some vegetables, you might use broccoli and cauliflower (both of which contain vitamin E). Note: Even though any frying tends to cause vitamin E to become unstable, you may be able to protect the E somewhat using this technique.

Summing Up the Antioxidant Way of Preparing Foods

To sum up, then, when you prepare your food, keep these principles in mind:

- Avoid wilted produce.
- Avoid excess trimming, cutting, chopping, slicing, washing, or soaking.
- Avoid excessive use of water in cooking.
- Avoid excessive use of heat in cooking. Long boiling or other extended cooking, or exposing foods to flame or smoke, as on a grill or open fire, can damage antioxidants. Those methods also may stimulate undesirable chemical changes in the food, as well as the production of extra free radicals.
- Try to consume the water you use in cooking—antioxidants will be present in it.

- Use syrup or liquids that result from thawing frozen foods.
- Do not refrigerate once-cooked foods more than one day.
- Store once-cooked foods in air-tight containers.
- Try not to reheat once-cooked plant dishes.
- Avoid keeping foods warm for more than thirty minutes before you serve.
- Do not buy pre-cut produce.
- Do not hold fresh produce in your refrigerator for more than a few days, and certainly not longer than a week. Buying frozen fruits and vegetables is a better choice if you think you won't consume the fresh produce within a few days of purchase.

Now let's put all this information together into an actual menu. The following one-day model meal plan, designed by the Nutrition Department at the Cooper Clinic, contains foods listed as carrying 1,120 mg of vitamin C; 36,085 IU of beta carotene; and 109 IU of vitamin E.

Our goal was to provide at least the basic daily antioxidant recommendation of 1,000 mg of vitamin C; 25,000 IU of beta carotene; and as much as possible of the recommended 400 IU of vitamin E. More than the minimum amounts of vitamin C and beta carotene have been included to allow for losses in these nutrients during food preparation. You will find that your foods have to be selected carefully to provide a daily diet containing 1,000 mg of vitamin C, but reaching a goal of 500 mg is not too difficult. Also, it is not hard to arrange your eating so that you get at least 25,000 IU of beta carotene per day. But to facilitate your food selections, you should try, as we did, to pick foods that contain both vitamin C and beta carotene.

Vitamin E presents more problems—primarily because to get high doses of vitamin E in your diet, you would also have to take in high amounts of fat. For example, you can always buy wheat germ oil in bottles in the vitamin section of many grocery stores. But while it would take twenty tablespoons of wheat germ oil to provide 400 IU of vitamin E, that amount would require you to consume 2,000 calories of fat! So in order to keep fat calories between 50 and 70 grams per day, a vitamin E supplement will be required.

This model menu contains just over 2,500 calories—an acceptable level for an active person who exercises regularly, but too much for most people on a weight-loss diet. If you want to lose weight on a lower-calorie regimen, or if you find you need to take in less food to maintain your current weight, you will have to reduce your food consumption, with an inevitable reduction in antioxidants. Taking the recommended antioxidant supplementation each day will allow you to reduce your calories and at the same time retain your free radical defense system.

The percentages of calories in this one-day plan come from these sources:

Protein	16 percent
Complex carbohydrates (fruits, vegetables, starches)	57 percent
Fat	23 percent (5 percent saturated fat; 10 percent polyunsaturated fat; 8 percent monounsaturated fat)
Sugar	4 percent

According to the American Heart Association and other national health organizations, your intake of calories from fat each day should be no more than 30 percent, with no more than 10 percent of those fat calories coming from saturated fats.

On the following pages you will find:

1. An outline of the three meals and two snacks in the model menu.
2. An explanation of the menu, with suggestions for recipes and preparation techniques.
3. A detailed analysis of the calories, fat, and antioxidants contained in each ingredient in the menu.

A One-Day Model Antioxidant Revolution Meal Plan

BREAKFAST
Orange Juice
Oatmeal Cereal and Wheat Germ
Raisins
Plain, Low-fat Yogurt
Fresh Strawberries

LUNCH
Cranberry Cocktail Juice
Whole Wheat Pita Pocket with
Steamed Broccoli and Cauliflower
and Part-Skim Mozzarella Cheese
Chilled Pasta Salad
Cherry Tomatoes

AFTERNOON SNACK
Fresh Kiwi Fruit

DINNER
Fresh Spinach and Tomato Salad
with Sunflower Seeds
Spicy Oil-free Vinaigrette Dressing
Stir-fried Chicken and Fresh Sweet Peppers
Steamed Carrots
Brown Rice
Chilled Fruit Salad

EVENING SNACK
Orange Juice
Bagel and Margarine

Menu: Recipes and Preparations

Breakfast
12 oz. orange juice
2 - 1 oz. pkg. Instant Quaker Oatmeal, *Extra Fortified*
1 T Raisins
8 oz. low-fat, plain yogurt
1-1/2 cups fresh strawberries

1. Prepare orange juice from frozen concentrate. This can be done the day before and stored in an airtight container in the refrigerator at 45 degrees Fahrenheit.
2. Prepare oatmeal according to the microwave instructions.
3. While hot, add wheat germ and raisins, and mix the ingredients together. Top with some cut strawberries, if desired.
4. Use whole fresh strawberries purchased within one to two days. Note: Rinse and remove stems from strawberries the day before. Pat dry and store in an airtight bag or container.
5. Mix some cut strawberries with the plain yogurt, if desired. If you add cut strawberries to the cereal or yogurt, cut immediately before serving.

Lunch
8 oz. cranberry cocktail juice (use bottled)
1 whole wheat pita pocket
1 cup steamed frozen broccoli and cauliflower medley
1 oz. grated part-skim mozzarella cheese
1 cup chilled pasta twists, cooked or prepared according to package directions
2 T Hellmann's Mayonnaise
Spices, salt-free
4 cherry tomatoes

1. Microwave frozen broccoli and cauliflower mixture according to the instructions. Season vegetables with desired spices.

2. Fill pita pocket with warm vegetables and top with grated mozzarella cheese.
3. Prepare pasta according to directions; refrigerate. Mix cold pasta with mayonnaise and season with spices if desired.
4. Use cherry tomatoes whole or cut in half immediately before you are ready to eat and mix with pasta salad.

Afternoon Snack
2 fresh, ripe kiwi

1. Select medium to large kiwi (one large kiwi is about 3-1/2 oz.).
2. Select plump, fragrant kiwi that yield to gentle pressure (like ripe peaches). If only firm ones are available, ripen at room temperature away from heat and direct sunlight for two to seven days. Speed the ripening process by placing kiwi in a paper bag with an apple, pear, or banana. Once ripe, store away from other fruits. Refrigerate ripe kiwi.
3. To peel, first cut off ends. Then peel with a sharp knife or vegetable peeler.

Dinner
1 cup fresh spinach leaves, torn into bite-sized pieces, refrigerated in airtight container
1/4 cup fresh chopped parsley, chopped, refrigerated in airtight bag
1/2 sweet green pepper, cut into 1/4- to 1/2-inch strips
1/2 sweet red pepper, cut into 1/4- to 1/2-inch strips
1/8 cup onion, cut into thin strips
1/2 cup fresh honeydew melon, peeled and chopped, chilled in airtight container
1/2 cup fresh cantaloupe, peeled and chopped, chilled in airtight container

1/2 cup brown rice

3 oz. chicken breast strips (skinless, boneless), cut into 1-inch strips

non-stick vegetable cooking spray

garlic powder

ground ginger

light soy sauce

1/2 cup steamed frozen carrots

2 tsp. Mazola margarine

1 fresh tomato

1 oz. sunflower seeds

2 T oil-free vinaigrette salad dressing

spices, salt-free

Fresh Spinach and Tomato Salad with Vinaigrette

If you are preparing the salad ahead of time, make sure you store the spinach leaves in an airtight container in the refrigerator.

1. Mix the spinach and sliced tomatoes together in a bowl.
2. Toss with vinaigrette, and sprinkle with sunflower seeds and salt-free spices.

Stir-fried Chicken with Peppers and Onions

1. Spray a skillet or wok with non-stick spray, and stir-fry the chicken over medium-high for a few minutes (until opaque).
2. Add the peppers, onions, garlic, and ginger to the skillet, and continue stir-frying until the vegetables are barely wilted. Add water as needed (2 T at a time) and a dash of light soy sauce. Cover and simmer three to five minutes longer.

Steamed Frozen Carrots

Microwave 1/2 cup frozen carrots according to package directions. Add 2 tsp. of Mazola margarine, and serve hot.

Brown Rice with Parsley

If you are preparing the parsley ahead of time, make sure you store chopped parsley in an airtight container in the refrigerator (to prevent oxidation).

1. Prepare the rice according to package directions.
2. Just before serving, stir in the chopped parsley.

Chilled Fruit Salad

If you are preparing the fruit salad ahead of time, make sure you store the cut fruit in an airtight container in the refrigerator.

1. Peel and cut honeydew melon into 1-inch cubes.
2. Peel and cut fresh cantaloupe into 1-inch cubes.
3. Combine the honeydew melon and cantaloupe in a large serving bowl and mix well.

Evening Snack

1-1/2 cups orange juice, made from frozen concentrate
1/2 bagel
1 T Mazola margarine

Step Three:
Estimate Your Total Daily Intake of Antioxidants—Including Foods and Supplements

The time has now arrived to estimate your total intake of antioxidants—a useful step to ensure that you are getting enough antioxidants each day, and also to identify the source of possible side effects. There are two ways to go about this: One is a simple approach for those who don't want to be bothered with any extra

Recipe Ingredients for SAMPLE DAY MENU

AMT	INGREDIENTS	CALORIES kcal	FAT gm	VIT C mg	B-CAROTENE IU	VIT E IU
1.5 c.	Orange Juice	168.0		145.5	291.0	
1.5 c.	Oatmeal, Inst. Quaker, Extra-Fortified	217.5	3.6	120.0*	10,230.0	60.0
0.25 c.	Wheat Germ, Honey Crunch Quaker	108.0	3.0			8.34
1.0 T	Raisins	27.1				
1.5 c.	Strawberry	67.5	0.8	127.0	61.5	0.27
8.0 oz.	Cranberry Cocktail	141.3		108.0		
1.0 pc.	Pita Bread, Whole Wheat	165.0	1.0			
1.0 c.	Broccoli & Cauliflower	35.0		44.0	2,950.0	
1.0 oz.	Mozz. Cheese, Reduced Fat	80.0	5.0			
2.0 medium	Kiwi, fresh	92.0	0.6	150.0	133.0	
3.0 oz.	Chicken Breast	140.4	3.0			
1.0 pc.	Sweet Pepper, cooked	20.0		170.0	280.0	
0.13 c.	Onion, fresh	6.0				
0.5 c.	Brown Rice	119.9	0.6			
1.0 c.	Spinach, raw	14.0		16.0	3,760.0	1.06

AMT	INGREDIENTS	CALORIES kcal	FAT gm	VIT C mg	B-CAROTENE IU	VIT E IU
1.0 pc.	Tomato	25.0		24.0	766.0	.42
0.5 c.	Carrots, ckd.	35.0			12,922.0	
2.0 tsp.	Margarine, Mazola	70.0	7.6		333.0	5.3
1 T	Vinegar	2.0				
1.5 c.	Orange Juice	168.0		145.5	291.0	
0.5 pc.	Bagel	81.5	0.7			
1.0 T	Margarine, Mazola	105.0	11.4		500.0	8.0
1.0 c.	Macaroni, ckd.	210.0	1.0			
0.8 pc.	Tomato	20.0		2.0	174.0	
0.5 pc.	Cantaloupe, fresh	28.5		34.0	2,579.0	0.11
0.5 c.	Honeydew, fresh	30.0		21.0	34.0	
1.0 c.	Yogurt, Nonfat, Plain	127.0				
1.0 oz.	Sunflower Seeds	175.2	16.1			14.18
1.0 T	Mayonnaise, Hellman's	100.0	11.0			11.0
0.5 tsp.	Soy Sauce, Lite	1.8				
0.25 c.	Parsley, Fresh	1.5		13.0	780.0	0.26
	TOTALS	2,582.2	65.4	1,120.0	36,084.5	108.94

*This amount is not included in total mg of vitamin C on the computer analysis.

Sources of nutrient breakdown:
Cooper Clinic computerized nutrient analysis.
J.A.T. Pennington, *Bowes and Church's Food Value of Portions Commonly Used*, 16th ed.
(Philadelphia: J.B. Lippincott, 1994).

focus on their food intake; the second is a more complex system for those who prefer a more detailed analysis of their diets.

The Simple Way

There are only four things you need to be concerned about with this approach. First, be sure that you are taking the minimum antioxidant supplements, as recommended in the previous chapter, for your age, sex, level, and body size. Second, be sure you have at least five to nine ample helpings of fruits and vegetables every day. Third, monitor yourself to be sure you aren't experiencing any of the side effects mentioned in this or the previous chapter. Fourth, notify your doctor if you are taking any other drugs or medications so that he can warn you about possible drug interactions.

Just by observing these four guidelines you can feel assured that you are taking in enough antioxidants without taking too many.

The Detailed Way

This approach involves more analysis of the actual amounts of antioxidants you are taking in through your supplements and diet. To evaluate your daily antioxidant intake, plan a one-day menu for yourself such as the model menu provided earlier in this chapter. Figure the amounts of each of the three main antioxidants—vitamin C, vitamin E, and beta carotene—you are getting in your foods.

Then, assuming you have taken all the precautions suggested to minimize the loss of nutrients in food preparation, reduce your estimate of dietary antioxidants by one-third. (You will recall that even with the greatest caution in food preparation, roughly a third of vitamin C and beta carotene may be lost during storage or cooking.)

Now, you should have a net dietary estimate for the amounts of vitamin C, beta carotene, and vitamin E that you consume in one day. Add those figures to the amounts of the vitamins you are taking in supplement form.

It is important for you to have this ballpark estimate for your intake of each antioxidant so that you can keep your consumption within the desired range. As mentioned in the previous chapter, there are almost never any side effects for those who stay within my recommendations. But just as a reminder, here are some discomforts or problems that a small number of people may experience when they begin to take larger doses of antioxidants.

Beta carotene. Levels above 30 to 40 mg per day (or 50,000 to 67,000 IU) can cause temporary yellowing of the skin. Also, in some animal studies, 30 mg a day (50,000 IU) taken with large amounts of alcohol have caused more liver damage than alcohol taken alone. Just to be safe, I recommend that beta carotene not be taken within two hours of any alcoholic drink, and that beta carotene never be taken if the daily consumption of alcohol exceeds more than four to six ounces.

Finally, heavy smokers should be aware that the National Cancer Institute's study of long-time male smokers in Finland revealed an 18 percent increase in lung cancer among the smokers taking large daily doses of beta carotene supplements.

Vitamin E. Supplements should not be taken if a patient is receiving anticoagulant therapy or has any condition contributing to impaired coagulation of the blood. (Vitamin E tends to block clotting of the blood.)

Controlled studies have reported few side effects in doses up to 3,200 mg (about the equivalent of 3,200 IU). Uncontrolled studies (those done without adequate scientific safeguards) and a few case reports have associated daily intake of more than 400 IU of vitamin E with occasional stomach or intestinal complaints, breast soreness, emotional upsets, fatigue, and decreased thyroid hormone levels.

The Journal of the American College of Nutrition concluded in 1992 that completely safe levels for practically everyone (except those on anticoagulant therapy) seem to be in the range of 200–400 IU per day. The majority of people can take higher doses, as

I have recommended for some groups, as long as they monitor possible side effects.

Vitamin C. A side effect at levels of 1,000 mg per day or higher may be diarrhea. Even at 500 mg per day, a few patients report frequent visits to the bathroom, abdominal pains, cramps, nausea, heartburn, headache, flushing, dry ears, nose, and throat, nosebleeds, or sleep disorders.

Very large doses of vitamin C may increase the potential for developing kidney stones because of the oxalate production resulting from the metabolic breakdown of vitamin C. Oxalate is a factor in kidney stone formation.

Finally, *Nutrition and the MD* reported in March 1993 that 1,500 mg a day may cause long-term risk of copper depletion in some patients if their dietary intake of copper is low. And chewable vitamin C tablets have been shown to cause damage to tooth enamel.

Again, such side effects occur in only a small number of people. But you should be aware of what they are and how your total intake of antioxidants may be contributing to them if you should begin to experience any of those difficulties. Many times, some side effects, such as the diarrhea or gastrointestinal problems that may be associated with vitamin C, will disappear after your body gets used to the higher intake. If the symptoms continue, you should reduce your intake of one or more of the antioxidants until the difficulties disappear.

To provide you with a practical tool to help you make maximum use of all this information, I have formulated the following chart. It is a "do-it-yourself" form, which you may copy in order to carry it around with you. There are spaces for you to fill in the amounts and times each day for taking your antioxidant cocktail. Also, there are lines for you to jot down your favorite antioxidant foods and the approximate vitamin content in them.

Note: You can buy the antioxidant supplements as part of a one-pill "antioxidant formula," which may have to be taken more

Your Personal Antioxidant Chart

Antioxidant Supplement	Daily Dosages	Time(s) Taken
Vitamin C	_____	_____
Vitamin E	_____	_____
Beta carotene	_____	_____
Other _____	_____	_____

Favorite Antioxidant Fruits	Approx. Antioxidant Content
_____	_____
_____	_____
_____	_____

Favorite Antioxidant Vegetables	Approx. Antioxidant Content
_____	_____
_____	_____
_____	_____

than once a day, since the amounts of the vitamins contained in each pill may be less than your daily recommendation. Or you can buy the supplements as separate pills (one vitamin E, one vitamin C, and one beta carotene). The supplements are available in most well-stocked supermarkets, pharmacies, or health food stores.

You now have three major defenses to ward off the damage from free radicals—lower-intensity exercise, antioxidant supplements, and a high-antioxidant diet. The final plate in your armor is a lifestyle that will allow you to reduce your personal exposure to the molecular outlaws.

8

Toward a Life Free of Free Radicals

T here is no way to retreat or escape to an environment that is totally free of free radical bombardment. Just living in modern society exposes us to harmful renegade molecules. Even if exposure to cigarette smoke could be completely eliminated, for example, air pollution would remain a significant problem.

Still, it is possible to minimize the threat that free radicals can pose to your health and thereby maximize your opportunity to live a long and productive life. As we have seen, lower-intensity exercise, antioxidant supplements, and the right kind of diet can strengthen your free radical defenses significantly. But to achieve a truly complete defense plan, it is also necessary to take certain evasive measures—which will enable you to avoid the presence of free radicals whenever possible. That means you must learn to identify and reduce or eliminate those forces in your environment that may continue to stimulate the production of free radicals in your body.

In many ways, this final step is the most difficult because it may require you to confront deeply ingrained personal habits, such as cigarette smoking. Or you may find that you have to overcome seemingly insurmountable obstacles, such as air pollution where you live or work. Furthermore, a great deal remains unknown about the impact of environmental factors on the body's production of free radicals. There is just not enough scientific data to enable anyone to say without reservation that such-and-such a level of pollution or electromagnetic fields will cause you to develop a certain health condition or disease.

These cautionary words are intended not to discourage you but to inject a note of reality into your quest for a radical-free way of life. It is impossible to live on this earth and escape from or eliminate all excess free radicals that may pose a threat. In fact, you will never even know the exact impact an environmental influence is having on your body. But you can take some steps to minimize your exposure to conditions and factors that may be linked to free-radical induced diseases. The more you think in terms of a lifetime defense plan against the threat of the molecular outlaws, the more likely it is that you will enjoy a longer, disease-free life.

Think in Terms of Your Whole Life

During the course of a normal life span, the typical human being is exposed to many environmental "triggers" that stimulate the production of free radicals in the body. Aware of the deadly impact that these unstable oxygen molecules can have on living tissues, researcher D. Harman, writing nearly four decades ago in the *Journal of Gerontology*, postulated what has come to be known as the "free radical theory of aging." Simply stated, the theory holds that the degenerative changes associated with the aging process may be caused by an accumulation of free radical damage.

Specifically, Harman suggested that some of the most destructive radicals and reactive oxygen species—including the superoxide, hydroxyl, and peroxide radicals—may be produced in excess over a normal life by exposure to cigarette smoke, ozone, car exhaust, ultraviolet light, and other such factors. The radicals generated by those forces may initiate various cellular and genetic mutations in the body, with a resulting loss of efficient enzyme production and cellular repair.

The extra damage over time to the cells through thousands of daily oxidative "hits" could lead to permanent damage in DNA; an appearance of aging through breakdown of tissues, such as those that make up the skin; and an increase in the incidence of chronic diseases, such as cancer and heart problems, which lead to premature debilitation and death.

This theory is still being tested and investigated, but increasingly, the evidence seems to support Harman's hypothesis—that free radical damage plays a role, and perhaps a significant role, in the aging process. In a 1993 report, *Medicine and Science in Sports and Medicine* by researcher Li Li Ji, for instance, a study of enzyme response in aging animals showed clearly that the antioxidant systems undergo significant alteration during aging—most likely as a result of oxidative stress.

The probable influence that free radicals have on aging presents some important practical questions.

- What are the major environmental "triggers" or danger areas that increase your free radical risk over a period of many years?
- Is there any way you can adjust your lifestyle right now to reduce your exposure to free radicals—and to the threat of premature aging?
- How can you design a personal lifetime strategy to free yourself as much as possible from the threat of free radicals?

The Major Free Radical "Triggers" that Put You at Risk for Premature Aging

When I evaluate my own life and the lives of my patients with an eye to reducing the threat of premature aging, I first concentrate on the basic parts of my Antioxidant Revolution program—the lower-intensity exercises, personalized supplement use, and a high antioxidant diet.

But then I begin to watch for other outside forces that are likely

173

to wear down the body's youthfulness and health over a period of many years; for example, practices, habits, and environmental factors that we are exposed to virtually every day and that are very difficult to avoid or eliminate. Any complete antioxidant program must take into account these influences that trigger the release of extra free radicals and heighten the risk of tissue damage and disease. You must first know your enemy before you can determine the best way to combat an attack.

Here are some of the major ongoing environmental triggers that put you at risk for oxidative stress and premature aging. (A reminder: In this list, I have omitted some of the topics already covered in depth, such as overtraining and excessive exercise, and food preparations that may reduce your free radical defenses.)

Trigger #1: Cigarette Smoke

If you smoke—or if you are exposed to secondary or "sidestream" smoke produced by other people's smoking at home, in the office, or elsewhere—the number of free radicals in your body will increase, and the likelihood that you will show early signs of aging will escalate.

Trigger #2: Air Pollution

Free radicals can increase in your body as you breathe ozone, car exhaust, and other pollutants in the air.

A 1993 study in the *New England Journal of Medicine* by D. W. Dockery and colleagues evaluated air pollution and mortality in six American cities and found that air pollution contributes to excess mortality. Air pollution was positively associated with death from lung cancer and cardiopulmonary disease.

Trigger #3: Inflammation

Free radical damage may be associated with inflammation of the muscles, ligaments, or joints. Inflammation may arise from physical injuries, including those suffered during sports, or long-term conditions, such as arthritis.

174

Trigger #4: Radiation

In addition to medical procedures utilizing radiation, such as x-rays and nuclear diagnostic scans, electromagnetic radiation is emanating all around us from high voltage wiring, computers, microwave ovens, electric blankets, and television. We often accept most of that radiation in our lives as benign; yet some of the earliest research on how free radicals damage the body focused on radiation.

Trigger #5: Sunlight and Other Ultraviolet Light

Exposure to sun, sunlamps, and other sources of ultraviolet light can stimulate the production of free radicals and increase the rate of aging—especially aging of the skin exposed to that light. In general, thirty minutes of exposure to sunlight each day gives you all that you need for your health, including the production of adequate vitamin D in your body. Any more sun than that is unnecessary and could lay the groundwork for skin cancer.

In advising my patients about how to avoid these triggers, I usually divide any action plan into two parts: (1) a simple, immediate response that I encourage them to begin on the very day that we talk; and (2) a longer-term, more comprehensive strategy for a lifetime.

What Should You Do Right Now to Reduce Your Exposure to Free Radicals?

Probably the most helpful advice that I can give you at this point is to tell you that it is not necessary to change your entire life *this moment* in an effort to get rid of all or most of the extra free radicals you may encounter. Instead, I suggest that today you pick one factor that you suspect is the most serious free radical trigger in your life, and begin to take steps to eliminate it—or at least do your best to limit your exposure.

For example, if you are a smoker, just tackle that problem.

Forget all the other free radicals that you may be exposed to. Anyone who has ever tried to quit smoking knows how difficult a task that can be. If you try simultaneously to eliminate smoking along with exposure to sunlight and several other free radical triggers, chances are that you will fail on all counts.

If you decide your main trigger is sidestream smoke, evaluate the places where you tend to be exposed to the smoke and try to avoid them. That can be as daunting a task as quitting smoking yourself—especially if you live with a smoker or work in an environment where many colleagues are smoking.

What can you do if you live or work with a smoker? It may be that all you can do as a first step is to limit other places where you may be exposed to smoking, such as the smoking section of restaurants. You may be able to spend only part of each evening in a chair or room occupied by the smoker in your family. Or you may be able to shift your desk or work space to a spot with vents, a location near an open window, or even a separate office or alcove, away from the smoke. The important thing is to identify the one important free radical trigger in your life and start to do something about it.

After you have begun to make progress in dealing with the single most important source of free radical activity in your daily life, you are ready to take the next step—the design of a more comprehensive radical-free strategy for the rest of your life.

A Strategy for a Life Free of Harmful Free Radicals

To formulate your radical-free—or at least radical-reduced—strategy, draw up a one-week schedule of the location or environment associated with each of your activities and involvements. It is best to write down everything you do and every place you go so that you will not miss any exposure that may place you at risk for free radical damage.

For example, your one-day diary might look like this:

6:30 A.M.—Arise, prepare for work.

7:30–8:15—Commute (car exhaust fumes are light to nonexistent for the first half of the drive, heavy for the last half; sun is in eyes all the way).

8:30–8:45—Coffee in office canteen (heavy cigarette smoking by fellow workers).

8:45–12:30 P.M.—Work in office (relatively smoke-free except for coffee break).

12:30–2:00—Lunch, often with business associates (about half the time in restaurants with poor ventilation).

2:00–5:30—Work in office (relatively smoke-free except for second coffee break).

5:30–6:45—Commute home (car exhaust fumes heavy for first half of drive, medium-heavy for last half).

6:30–7:15—Jog, outdoor calisthenics (can smell car exhaust fumes on roads during exercise).

7:30–11:00—Dinner, reading, television at home (spouse smokes for about an hour in the family room).

After you have listed these activities and the environment in which they are conducted, you will be in a better position to analyze your situation and make some adjustments.

For example, you may decide that you need to wear a hat, don sunglasses, or otherwise cover yourself up during your commute, when you are exposed to the sun. To reduce your exposure to air pollution, you might try going to work earlier in the morning or returning later at night. Avoiding the smoking of fellow workers or that in restaurants during the day may often be possible just by choosing a different place to eat or snack.

You will probably find that if you shift your jogging to before dawn or after sundown, the air pollution will be less. Do not jog or exercise within thirty feet of a busy highway, particularly during daylight hours. Also, you can most likely reroute your exercise path to avoid heavy traffic or polluted areas.

Finally, you cannot move out of your home just because your spouse smokes. If the smoking only goes on for an hour, however, you may be able to shift your activities to a different room during that period. Or, as many families have done, you could encourage the smoking member to limit smoking to one room, or even better, to go outside to smoke.

The idea with these adjustments is not to become obnoxious or to transform yourself into a holier-than-thou "health nut" in your dealings with colleagues or family members. Rather, some quiet but decisive shifts in the way you conduct your activities can go a long way toward reducing your exposure to the free radicals that may reduce immunity, cause disease, and promote premature aging.

The Antioxidant Revolution, when considered in its many aspects, is indeed an all-encompassing, truly revolutionary approach to health and longevity. At the same time, most of the changes that I am suggesting can be achieved with a minimum of disruption in your exercise, eating, and living habits. All that is required is that you take the free radical threat seriously—and respond reasonably to a movement that is fast becoming part of the scientific mainstream of good health and medical practice.

Epilogue

The antioxidant/free radical story sometimes reminds me of a breaking news event. As soon as you think you have described all the important elements of an issue, something new pops up. To get a taste of how this exciting topic is always changing, just look for references to antioxidant research in your local newspaper, listen to television magazine features—or, as I often do, watch for reports of scientific seminars and workshops that focus on the topic.

Just as I was putting the finishing touches on this manuscript, I was alerted to a workshop on free radicals and antioxidants that was held August 31 through September 1, 1993, in Washington, D.C. Leading scientists in the field explored the impact of food and non-food (supplements) antioxidants on health. The specific purposes of the meeting included these three:

1. To evaluate critically the existing data showing a beneficial effect on human health of antioxidants found in food and food supplements.
2. To develop a scientific approach to be used in future studies which would clearly show the impact of antioxidants on human health.
3. To identify a way to increase their understanding of the role that these compounds might play in promoting our health; they also desired to develop an effective way to communicate that information to the public.

During the workshop, representatives of the U.S. Food and Drug Administration (FDA) outlined some fundamental principles that will guide the agency in evaluating potential new claims, particularly as those claims might relate to the use of antioxidant vitamin and mineral supplements. Among other things, the FDA

179

said that approval of marketing the supplements should be based on such factors as:

- The scientific truth of any claims.
- A consideration of both potential risks and benefits.
- The involvement of the scientific community in the supplement industry through advisory councils, workshops, and other mechanisms.

The FDA also pledged to work hard to take full advantage of the agency's major vehicle to communicate information to the consumer—the food label.

Many of the scientists at the Washington workshop shared their understanding of the latest facts, findings, and trends in antioxidant and free radical research. Below are the names of some of the leading participants and a brief summary of their reports:

Dr. Barry Halliwell of the University of California at Davis proposed that an antioxidant be defined as any substance that protects tissues from oxidative damage. He reminded participants that reactive oxygen species play a beneficial role in maintaining health since white blood cells use them to kill invading bacteria.

But free radicals can cause serious oxidative damage unless they are neutralized or inactivated. Unfortunately, the body's defense system against free radicals is not 100 percent effective, and free radicals are potentially adverse factors in a wide range of chronic diseases. Furthermore, it is not always clear whether free radicals are the cause or the result of tissue damage. Yet we do know, Halliwell said, that such damage to the body's tissues can cause further free radical production and lead to additional tissue damage.

Dr. Lawrence Machlin of Hoffman-Laroche, Inc., concluded that epidemiological studies (those of large population groups) consistently suggest a positive effect of antioxidants on cancers of the lung, stomach, esophagus, and throat. The links are strongest for beta carotene (or vitamin A) and lung cancer on the one hand, and vitamin C and stomach cancer on the other. He noted that there is less conclusive data relative to heart disease, but use of vitamin E supplements has been associated with decreased risk in both males and females, and supplementation with beta carotene has reduced the incidence of cardiovascular problems in a population of people with clinical signs and symptoms of heart disease.

Dr. Thomas Slaga of the University of Texas Anderson Cancer Center reported that the evidence for a protective effect of antioxidants is strongest at the very early stages of a disease, such as cancer. There is also the possibility that vitamin C and beta carotene may be protective against cancer of the stomach in humans through a mechanism unrelated to their antioxidant properties. In other words, increasing the consumption of foods rich in vitamin C and beta carotene, such as fresh fruits and vegetables, may provide some protection from cancer due to other, as yet unknown ingredients. These ingredients include "phytochemicals," which are compounds in plants that are believed to have health-enhancing effects.

Dr. David Janero, staff scientist, Ciba-Geigy Corporation, said there seems to be little question about the value of antioxidants in reducing the risk of ischemia-reperfusion injury. You will recall that that involves the temporary loss of blood and oxygen to body tissues (especially the heart during a heart attack), and then the rushing of blood back into the deprived tissues. Without antioxidant therapy, this phenomenon can be lethal within seconds after circulation is restored to the heart.

Dr. Balz Frei of the Harvard School of Public Health stated his support for the role that antioxidants play in controlling or reducing the atherosclerotic process. He affirmed the position that the free radical oxidation of the "bad" LDL cholesterol particle is the process that initiates the atherosclerotic process (i.e., clogging of the arteries). Furthermore, he noted that experiments involving vitamin C in a laboratory setting (in vitro) were able to protect the LDL cholesterol completely from free radical damage.

Considerable protection has also been provided by other antioxidants, including vitamin E, Dr. Frei said. In their human (in vivo) studies, Drs. Ishwarlal Jialal and Scott Grundy have confirmed this observation and labeled vitamin E as the "superior antioxidant" for protection against LDL oxidation.

D. Shambhu Varma, Director of Ophthalmology Research, University of Maryland School of Medicine, presented data suggesting that vitamin C, vitamin E, and certain "bioflavonoids" (substances obtained from plants like buckwheat and horse chestnut bar) can protect against cataracts.

Dr. Priscilla Clarkson, Department of Exercise Science, University of Massachusetts, discussed the current thinking about exercise-induced free radicals and their potentially harmful effects. She reported that physical conditioning does increase the body's antioxidant defense systems by increasing circulating antioxidants. Consequently, up to a point, fitness-producing exercise can reduce the oxidative damage that can be caused by intense exercise. Furthermore, individuals who exercise strenuously and exceed the body's natural defenses can reduce the extent of the damage by taking antioxidant supplements. (Her observations and findings further confirm the information and recommendations presented in chapters 4 and 5.)

Dr. Clarkson also reminded the group that even though antioxidant supplements do not enhance physical performance at sea level, studies have shown some enhancement of athletic ability among mountain climbers at higher altitudes.

In conclusion, the participants in this workshop urged that the potential health benefits of increased antioxidant intake should be reasonably balanced with the need for more scientific studies and validation. They stated that since cancer and heart disease are the two leading causes of death in the United States, close attention should be paid to any possibility of reducing those deaths in an effective, reasonably safe, and inexpensive way.

Finally, just as this book was going to press, the National Cancer Institute reported in April 1994 in *The New England Journal of Medicine* the results of a five-to-eight-year study on more than 29,000 long-time male smokers in Finland. The smokers were divided into four groups, with the first group getting daily beta carotene supplements; the second, daily vitamin E supplements; the third, both beta carotene and vitamin E; and the fourth, no supplements. The results of this study have been summarized in chapter 2 and elsewhere in the text, as well as in appendix 6. That is the latest word on antioxidants and free radicals—from my desk to yours. Stay tuned for further developments as additional significant breakthroughs occur in the exciting and life-giving Antioxidant Revolution.

1

The Language of
the Antioxidant Revolution

To grasp the true nature and enormity of the free radical threat, it is important to understand the scientific terminology that is used to describe the specific molecular outlaws that are now invading your body. Also, you need to know the names and characteristics of your antioxidant defense systems. Armed with this "language" of the Antioxidant Revolution, you will be in a position to understand more fully the defenses available to you and the practical ways to strengthen those defenses.

Who Is Really the Enemy?

When we think of oxygen, we usually associate it with breathing and air, the very sustenance of life. Without oxygen, we could not live, even for a short time. We are all "aerobes" in the sense that we live in air, with our atmosphere made up of almost 21 percent oxygen.

But not all oxygen molecules are alike. Most of the oxygen we breathe is stable and essential to our health and well-being. Other unstable oxygen molecules—which include the free radicals and their kin—may also behave in ways that are good for us. But on occasion, the radicals may turn into a classic case of good guys gone bad.

Your body desperately needs the services of the well-organized, highly disciplined force of powerful molecules known as "reactive oxygen species." They have been given that name because each of them represents a volatile and aggressive variety of the oxygen molecule. They also go by the name "oxidants" because they have the power to combine with, or "oxidize," other molecules.

Four particularly important molecules reside under the reactive oxygen species umbrella. They include the hydroxyl radical and the superoxide radical—both of which are known as "free radicals." Also, two related molecules, the oxygen singlet and hydrogen peroxide, are known as "non-radical" variations of the reactive oxygen species.

The radical and non-radical reactive oxygen species differ in their molecular structures—a fact that affects their modus operandi in your body. Free radicals have one or more unpaired electrons. That structure is relatively unusual because most electrons in molecules come in pairs. As these single electrons move about in their orbits, they are inherently unstable. The reason: Stability on the atomic level depends on having one electron balanced by another electron in each orbit.

As a result of their instability, free radicals are constantly on the lookout for other molecules they can lock on to, like little magnets. (At times, radicals are also referred to as "cellular wrecking crews" because of the damage they inflict after they combine with another molecule.) They exist alone for only a fleeting microsecond before they smash into another molecule.

The other members of the reactive oxygen species, the non-radicals, are constructed differently from free radicals. Non-radicals all have paired electrons, so they tend to be somewhat more stable than their free radical brethren. But they are still more active than most other molecules. Consequently, they are well suited for their demanding but beneficial work in your body.

What exactly is this good work that they do? When they are under control, the reactive oxygen species, as a group, are truly all-purpose utility players in keeping your physical systems working properly. Here is an overview of where they come from and how they operate.

Free radicals and non-radical reactive oxygen species arise from a variety of locations and situations inside your body. They may emanate from:

- *Your body's basic metabolism.* When the food you eat is transformed into energy by the energy factories (mitochondria) in your cells, free radicals blast out during the metabolic process. If things are working correctly, one of the most common of these radicals, the superoxide radical, is immediately broken down by an enzyme called "superoxide dismutase" (also known as SOD).
- *White blood cells.* When dangerous bacteria enter your body and threaten you with infectious disease, different types of white blood cells—such as "leukocytes," "monocytes," and "macrophages"—begin to throw out free radicals to destroy the intruders. You might think of these radicals as minuscule missiles that zero in on your body's internal enemies.
- *The lining of the blood vessel walls.* This lining, known as the "endothelium," puts out the superoxide radical to help regulate the contraction of the smooth muscles of the blood vessels. Controlling the tone of the lining of the vessels is essential to maintaining proper blood flow.
- *The red blood cells.* An iron-linked protein in the hemoglobin of the red blood cells uses reactive oxygen species to break down and utilize oxygen in the body.

As you can see, free radicals and their non-radical kin are constantly busy throughout your body, keeping your different systems going and protecting you from outside attack. Unfortunately, they can also be transformed into turncoats and end up directing their formidable strengths against you.

The Making of a Renegade

Although we could not function without the good guys—the reactive oxygen species that are under control and doing their jobs—we can also be placed in deadly danger by those that begin to break the body's laws.

What causes so many of these molecules to become outlaws? Our bodies have been designed to operate at their best as part of a delicate balancing act, with just enough radical and non-radical reactive oxygen molecules to keep us in good health. When that balance is upset, the trouble begins. Unfortunately, we live in a world that is full of hostile forces able to trigger the release of too many radicals in our bodies.

Some of the factors that may put you in jeopardy are:

- *Cigarette Smoke*
  Some radicals, which are actually emitted by the smoke itself, enter the body through the respiratory system. These unstable oxygen molecules may damage tissue in the lungs directly, or they may trigger the release of reactive oxygen species by the body's cells, including the white blood cells.
- *Air Pollution*
  This trigger works much the same way as cigarette smoke, with some radicals coming directly into your body and others being produced as the polluted air contacts cells.
- *Certain Drugs*
  These include anticancer drugs, such as doxorubicin (Adrimycin). This drug, in addition to fighting the cancer, may cause heart problems through the release of radicals.
- *Ultraviolet Light*
  These powerful rays, which are emitted by the sun or a sun lamp, can cause cell damage.
- *Pesticides and Other Chemical Contaminants*
  You constantly ingest these substances through the foods you eat and beverages you drink.

- *Excessive Exercise*
  Any overtraining can produce extra free radicals, but the most serious damage tends to be produced by "ultra" type training and competitions (such as ultra-marathons, which involve racing longer distances than the traditional marathon).

- *Joint and Tissue Injuries, Including Sore or Strained Muscles*
  The trauma and inflammation produced by injuries during exercise or an unaccustomed, heavy workout, commonly result in excess radicals. They are emitted at the site of the injury or in the overused muscle.

- *Uncontrolled Diabetes*
  Years of exposure of diabetic patients to high levels of blood sugar can engender unstable oxygen molecules.

- *Radiation*
  Too many free radicals may appear in the body when you are exposed to x-rays or any radioisotopic device, such as a CAT or P.E.T. scan.

- *Emotional Stress*
  The production of extra free radicals when you are under excessive pressure may be the reason stress is associated in many studies with heart attacks and cancer.

- *Asbestos and Related Fibers*
  This solid "particulate" matter may enter the lungs through breathing. The fibers then damage white blood cells, and finally trigger the release of highly destructive free radicals.

- *Reperfusion Injuries*
  "Reperfusion" refers to the return of blood to an organ or tissue that has been temporarily deprived of blood. This occurs after a non-fatal heart attack or a stroke when blood has been cut off to the heart or brain for a short period. Bursts of free radicals may be released during this process. At times, rapid reperfusion may lead to fatal interruption in the rhythmic beat of the heart.

Reperfusion injuries may also occur when organs or tissues are temporarily deprived of blood by the use of a tourniquet or a bypass pump during heart or coronary artery surgery.

- *Shunting of Blood to the Digestive Tract After a Heavy Meal*

Blood may be diverted from the muscles, heart, and even the brain to the stomach and intestines after you eat a big meal. This process, which involves a kind of reperfusion, is what makes you sleepy after a meal—and may cause heart patients to have chest pains. There may even be muscle cramping and nausea if exercise is performed immediately after eating. It is likely that the release of free radicals is a major influence behind these feelings of discomfort or illness.

When such factors trigger the release of too many radical and non-radical reactive oxygen species, the results can be disastrous.

First of all, the outlaw molecules may play a central role in atherosclerosis, or hardening of the arteries. Many experts believe that the build-up of artery-clogging plaque in the vessels is directly related to the oxidation of the "bad" or LDL cholesterol by the radicals. What happens to your LDL cholesterol is similar to what happens to food when it spoils or goes rancid after being left out on the counter, where it is exposed to oxygen and heat.

The renegade molecules may also injure the nucleus of a cell and break the genetic chains of DNA. This damage can cause various cancers. The outlaws may otherwise ravage the membrane and other cellular structures. Possible diseases include cataracts, arthritis, infant blindness (caused by exposure of premature infants to an excessively high oxygen atmosphere in incubators), premature aging, and impaired immunity.

All four of the main members of the reactive oxygen species get involved in this feast of destruction—with the fastest, most active, and potentially the most deadly of all being the hydroxyl radical. Furthermore, these outlaws don't exist or operate independently.

Typically, they are the product of chain reactions, where one of the renegade molecules gives rise to another.

For example, hydrogen peroxide may produce an oxygen singlet. Or chemical interactions involving hydrogen peroxide may lead to the appearance of a hydroxyl radical. One radical explosion often leads to another, and another, and another. And along the way, more and more damage is inflicted upon your body. By living and breathing—simply by existing in the world today—you are exposed to poisons, events, and people whose habits place you at serious risk from the molecular outlaws. What can you do to protect yourself? Is there no force for good that can rescue you from the deterioration that threatens your body?

Fortunately, there are two such positive forces that can turn aside the radical threat and put you on the track to a healthier and longer life. The first is a kind of police force that exists in your body—but may need some strengthening. The second is an external set of allies—a kind of antiradical Marine Corps that is just waiting for the signal from you to launch a counterattack. Together, these two forces compose the foundation of the Antioxidant Revolution that can enable you to regain control over your health and life.

Your Body's Police Force

If it weren't for the environmental and lifestyle factors that help produce the extra radicals in your body, you would probably possess enough internal "policemen" to keep your reactive oxygen species under control. These natural defenses are called "endogenous antioxidants."

The word "endogenous" means that they are produced inside your body. And "antioxidant" means that they have the power to disarm or neutralize the radical and non-radical reactive oxygen species. (The term "oxidant" is sometimes used to refer to the reactive oxygen species because they readily combine with, or oxidize, other molecules.)

There are three main endogenous antioxidants: First, there is super oxide dismutase (or "SOD," as it is commonly known). This particular policeman has the power to convert the superoxide radical into hydrogen peroxide, which in turn, through a separate process, is turned into water and oxygen.

The second internal antioxidant is catalase, which removes the hydrogen peroxide. Through this action, catalase helps prevent cell damage, such as the breaking of DNA strands, which may eventually lead to cancer.

The third naturally produced antioxidant is glutathione peroxidase (or GSH), which is even more important than catalase in removing hydrogen peroxide.

Overall, these three naturally produced antioxidants, along with other enzymes and molecules in the body, do a good job of maintaining some control over the reactive oxygen species. But with the prevalent environmental triggers and other factors that release extra radicals nowadays, the internal police force just isn't adequate.

It is possible to strengthen the endogenous antioxidants to some extent—primarily by using gradual, lower-intensity training techniques to increase your physical fitness. This sort of program can actually reduce the tendency of your body to produce excessive radicals during intense exercise and can also bolster your natural antioxidant defenses. (See chapter 8.)

Regardless of how strong your internal police force is, however, or how much you are able to strengthen it through exercise training, you still need outside help. That assistance is available mainly in the form of three vitamins that I call the "Antioxidant Marine Corps."

Send in the Antioxidant Marines!

To acquire full protection from the reactive oxygen species that are running amok in your system, it is absolutely necessary that

191

you fortify your body's natural defenses with "exogenous antioxidants." These free radical scavengers, which you can send in from the outside, include vitamin E, vitamin C, and beta carotene, a "carotenoid" that is a precursor of vitamin A.

Each of these vitamins is explored in detail in chapter 6, which talks about supplements, and in chapter 7 on nutrition. But here is a brief overview.

Vitamin E

This vitamin serves many important functions in protecting your body from free radicals. It exists naturally in LDL and "takes the shots" of the free radicals that are darting about. By sacrificing itself to the free radicals, vitamin E keeps the LDL from being oxidized, a process that ultimately ends up in "foam cells." This protective action is extremely important because the foam cells, which are packed with oxidized LDL cholesterol, will eventually group together to form plaque and clog the arteries. Remember, this development of plaque is a major step in atherosclerosis and heart attacks.

Unfortunately, the naturally produced vitamin E stores are often inadequate, so some outside help is usually necessary. Taking extra vitamin E into the body through diet and supplementation will strengthen your natural defenses and provide a strong shield against the outlaw oxygen molecules.

Vitamin E has also been linked to protection against gastrointestinal cancer and lung cancer. This anticancer defense is based on the ability of vitamin E to stop renegade radicals before they pierce deep into a cell and damage the nucleus and DNA, or genetic code. Cells with damaged DNA are more likely to become malignant.

Vitamin C

This vitamin enhances the effect of vitamin E in preventing the oxidation of LDL cholesterol and the formation of plaque. Also, vitamin C has been associated with the prevention of cancers of the esophagus, larynx, oral cavity, pancreas, stomach, rectum, and cervix.

Furthermore, vitamin C acts as a nitrite scavenger to help clean up the residues of cigarette smoke. It has even been shown to have a protective effect against injury to the DNA in human sperm. Such damage can affect sperm quality and lead to genetic defects.

Beta Carotene

Beta carotene has the power to provide some protection from lung cancer, oral cancer, bladder cancer, and rectal cancer. It has also been linked to the prevention of skin tumors, which may be triggered by exposure to ultraviolet light.

You have now been introduced to the main front line operators in the Antioxidant Revolution:

1. The "good guys"—or the properly functioning types of reactive oxygen species (including free radicals and non-radicals). They do essential, beneficial work in your body, such as protecting you from invading bacteria.
2. The molecular outlaws. These are members of the reactive oxygen species that have gone out of control and have begun to damage and destroy organs and tissues.
3. Your body's internal police force. These peacekeepers include the three main "endogenous antioxidants" that neutralize the extra radicals and prevent injury to your body.
4. The essential outside troops, or "marines." Three main vitamins—E, C, and beta carotene—provide the internal antioxidants with the help they need to keep you healthy.

2

The Scientific Foundations for the Antioxidant Revolution

M uch has already been said in the main text of this book about the scientific underpinnings on which I base my Antioxidant Revolution. But I believe it is also important, for the benefit of those interested in more technical explanations, to go into greater detail about some of the studies and findings that have led me to my conclusions.

This discussion will highlight scientific reports that link free radical damage and the antioxidant response to heart and cardiovascular disease, cancer, cataracts, and a number of other diseases. (Complete citation of scientific journal articles referred to in this appendix may be found in references listed for chapters 1 and 2.)

The Internal Plot Against Your Heart and Blood Vessels

In a symposium published on February 25, 1993, in *The American Journal of Cardiology*, guest editor Dr. Antonio M. Gotto summarized the current knowledge of antioxidants and lipid (blood fat) metabolism this way:

There is increasing interest in the potential use of antioxidants as therapy for atherosclerotic disease, in particular discovery of the role played by oxidation of low density lipoprotein (LDL) in the development of atherosclerosis in the Wantanabe rabbit. Results from the Physicians' Health Study have suggested that beta-carotene may be beneficial in reducing risk of cardiovascular events, as beta-carotene is a well-recognized antioxidant in the LDL fraction. Some case-controlled studies have reported decreased levels of ascorbic acid (vitamin C) and vitamin E in patients with coronary artery disease.

One of the articles in this symposium focused on the effect of probucol, a drug that has antioxidative, as well as lipid lowering, effects. Researcher Goran Walldius and colleagues from the King Gustaf V Research Institute, Karolinska Institute, Stockholm, Sweden, found that probucol not only reduced cholesterol levels, but also reduced the formation of TBARS. The researchers noted that this TBARS result indicated that "probucol protected LDL from oxidation."

In another recent report in *Clinical Cardiology*, Dr. William S. Harris concluded:

> Nutrients with antioxidant properties such as vitamins C and E, beta carotene, and monosaturated fatty acids (when they replace polyunsaturated fatty acids) can reduce the susceptibility of LDL to oxidation. Antioxidant therapy, if proven useful, should be considered an adjunct to lipid-lowering therapy in order to have the greatest impact on coronary heart disease.

What is the future of research into the effect of antioxidants on cardiovascular disease? Here are some "summary recommendations" from proceedings at a workshop on "Antioxidants in the Prevention of Human Atherosclerosis" during meetings of the National Heart, Lung, and Blood Institute on September 5–6, 1991:

- More basic research is needed on the mechanisms involved in the oxidative modification of LDL and on the ways in which various antioxidants influence it.

- More animal studies are needed to establish firmly that protection of LDL against oxidative modification does influence the progression of lesions.
- More research is needed on the possible effects of antioxidants on the later stages of atherosclerosis and on thrombosis.
- It was the consensus that the evidence available justifies a clinical trial of natural antioxidants (associated with no increase in risk) but that clinical trials with antioxidant drugs (which might carry deleterious side effects) should be deferred until more knowledge is available.

In other words, the experts at this workshop—which included the preeminent antioxidant and free radical researcher Daniel Steinberg of the University of California in San Diego, as well as my consultant, Dr. Scott M. Grundy—gave the go-ahead to further research with natural antioxidants, such as vitamins E, C, and beta carotene. But they cautioned against further research at this time on drugs, such as probucol.

How Antioxidants Combat Cancer

Gladys Block, Professor of the School of Public Health, University of California, Berkeley, sums up the current understanding this way in the July 1992 issue of *Nutrition Reviews:* "I believe the data clearly support approval of a health claim of the general form: 'A diet high in antioxidant vitamins may help reduce the risk of some cancers.' This is exactly parallel to what the FDA is proposing to approve with regard to health claims about dietary fat and cancer."

In support of that position, Tim Byers and Geraldine Perry concluded in a 1992 article in the *Annual Review of Nutrition:* "Antioxidant micronutrients, especially carotenes, vitamin C, and vitamin E appear to play many important roles in protecting the body against cancer. They block the formation of chemical carcinogens in the stomach, protect DNA and lipid membranes from oxidative damage, and enhance immune function."

But Byers and Perry make clear that much more work needs to be done in this area:

> Why do micronutrient effects differ by organ site, what are the optimal doses at which risk is reduced, and are there potential adverse effects? [These] are questions that need to be answered. . . . Although many important questions remain . . . there is a strong scientific basis for current U.S. recommendations that emphasize frequent fruit and vegetable consumption. Fruits and vegetables seem to protect the human body against cancer, perhaps because they protect the body against oxidative damage.

Antioxidant treatment has also been found to be helpful in conjunction with chemotherapy and irradiation. Here is what Researcher K. Jaakkola and colleagues reported in *Anticancer Research* in 1992:

> Antioxidant treatment, in combination with chemotherapy and irradiation, prolonged the survival time of patients with small cell lung cancer. . . . We also noticed that the patients receiving antioxidants were able to tolerate chemotherapy and radiation treatment well. Surviving patients started antioxidant treatment in general earlier than those who succumbed.

Finally, epidemiologic studies involving large populations are showing that high blood levels of antioxidants are associated with protection against various cancers. Researcher J. Chen and colleagues reported these results in 1992 in the *International Journal of Epidemiology* from a study of men and women in sixty-five rural counties in the People's Republic of China:

> Plasma levels of dietary antioxidants were consistently negatively correlated with cancer mortality rate. Ascorbic acid (vitamin C)

was most strongly negatively associated with most cancers and sele-nium with oesophageal and stomach cancers. Beta carotene was found to have a protective effect independent of retinol, particu-larly for stomach cancer.

Your Vulnerable Eyes

Free radicals may damage your eyes—specifically through cata-racts, which involve the clouding of the lens. S. D. Varma of the Department of Ophthalmology, University of Maryland School of Medicine in Baltimore, stated the problem—and the possible solution—this way in the 1991 issue of the *American Journal of Clinical Nutrition:*

Studies . . . indicate that the ocular lens is physiologically damaged when exposed to an environment of active species of oxygen, com-monly referred to as oxyradicals. Several photochemical and non-photochemical models have been described. The results suggest that an intraocular generation of active oxygen may constitute a significant risk factor in the overall pathogenesis of senile cataracts. The cataractogenic effect of oxyradicals, however, can be thwarted by nutritional and metabolic antioxidants such as ascorbate (vita-min C), vitamin E, and pyruvate. These agents, therefore, may be useful for prophylaxis or therapy against cataracts.

In support of Varma's findings, James Robertson and colleagues reported the following investigation in the *American Journal of Clinical Nutrition* in 1991:

A recent epidemiologic study found that cataract patients tended to have lower serum levels of vitamins C, E, or carotenoids than did control subjects. The present investigation, which compared the self-reported consumption of supplementary vitamins by 175 cataract patients with that of 175 individually matched, cataract-free subjects, revealed that the latter group used significantly more supplementary vitamins C and E. . . . Because their results suggested a reduction in the risk of cataracts of at least 50 percent, a randomized, controlled trial of vitamin supplementation in cataract prevention may be warranted.

The Free Radical Involvement in Other Diseases—and the Antioxidant Response

Free radicals have been implicated in more than fifty diseases. We have already considered some of the major areas, including cardiovascular problems, different cancers, and cataracts. Here are some of the other potential sources of free radical trouble— and possible antioxidant responses.

Parkinson's Disease

Stanley Fahn of the Department of Neurology, Columbia University College of Physicians and Surgeons, and the Neurological Institute of New York, Presbyterian Hospital, New York, reported his study in 1991.

He noted that high dosages of tocopherol (vitamin E) and ascorbate (vitamin C) were administered to patients with early Parkinson's disease. The primary final treatment was to give the patients Levodopa, a drug used to control Parkinson's. In the study, he compared the time that the Levodopa became necessary in patients getting the antioxidants with the time the drug became necessary with patients who didn't get the vitamins.

His conclusion: "The time when Levodopa became necessary was extended by 2.5 years in the group taking antioxidants. The results of this pilot study suggest that the progression of Parkinson's disease may be slowed by the administration of these antioxidants."

Although at this point further studies have not confirmed Fahn's preliminary findings, his results certainly warrant future investigations into a possible antioxidant-Parkinson's connection.

Oral Cancer

At the University of Arizona and Veteran Administration Medical Center in Tucson, researcher Harinder S. Garewal reported in 1991 in the *American Journal of Clinical Nutrition:* "Recent data suggest that retinoids and carotenoids may be effective in reversing a putative 'field cancerization' defect in the epithelium at risk for oral cancer."

He concluded that because of its lack of toxicity and side effects, "Beta carotene is a very attractive agent for chemoprevention. It suppresses micronuclei in exfoliated oral mucosal cells from subjects at risk for oral cancer and recently has been shown to be active in reversing leukoplakia." (Leukoplakia is a white patch of oral mucous membrane that cannot be wiped off, but also cannot be diagnosed as a specific disease—though it may be a precursor to skin cancer.)

In a recent study by Hans Stich and other researchers at the Environmental Carcinogenesis Unit, British Columbia Cancer Research Center, Vancouver, BC, Canada, beta carotene and vitamin A were given to tobacco-chewers who had leukoplakia. The beta carotene caused remission of the leukoplakia in nearly 15 percent of the subjects.

Furthermore, the researchers discovered that the "protective effect of the original treatment could be maintained for at least eight additional months by administration of lower doses of vitamin A or beta carotene," according to a 1991 report in the *American Journal of Clinical Nutrition*.

Brain Hemorrhage in Premature Babies

Researcher Malcolm Chiswick and colleagues of the Neonatal Medical Unit and Department of Radiology, North Western Regional Perinatal Centre, Saint Mary's Hospital, Manchester, United Kingdom, administered vitamin E to preterm babies, born before thirty-three weeks of gestation.

A high proportion of newborns in this category develop bleeding in and around the lateral ventricles of the brain. Yet the administration of vitamin E by this group of scientists resulted in a significantly lower incidence of intraventricular hemorrhage than a control group.

The researchers concluded in the *American Journal of Clinical Nutrition* in 1991 that "vitamin E protects against IVH (intraventricular hemorrhage) in preterm babies."

Reperfusion Arrhythmias During Bypass Surgery

When the heart muscle has been deprived of blood and oxygen during bypass surgery, rapid restoration of circulation at the conclusion of the surgery may be associated with a severe and at times fatal heart irregularity. Studies in which blood transfused during surgery has been enriched with vitamin E have shown a marked reduction in the frequency of reperfusion rhythm irregularities. It is very likely that this beneficial effect is the result of the vitamin E counteracting the free radical damage which is responsible for the severe rhythm disturbances.

Obviously some of these studies, such as the ones involving Parkinson's disease and cataracts, are preliminary, and much more work needs to be done to find exactly how antioxidant therapy may be used to counter free radical damage. But many of the findings, including those relating to cardiovascular disease and many forms of cancer, simply cannot be ignored either by practicing physicians or patients. It is virtually certain that future studies will confirm many of the benefits of antioxidants that are now in a very preliminary state of understanding.

In the meantime, I am convinced that you can't go wrong by embarking on a serious antioxidant program. In fact, the chances are very good that when you bolster your antioxidant defenses, you will be protecting yourself against many more health problems than we can even imagine at this early stage of scientific investigation.

The Benefits of Exercising in the Higher Ranges of a Fitness Program

Many people who embark on an exercise program quickly become dissatisfied with minimal activity and begin to increase the intensity of their workouts. Furthermore, the American Heart Association, the Surgeon General of the United States, and other health authorities have encouraged aerobic fitness that exceeds the lower-intensity, fifteen points per week program presented in this book.

What are the causes for that pressure to increase your level of activity? One reason is that experimental studies have shown that reaching a high level of fitness can increase certain benefits, including the following:

- An improvement in blood lipid profiles, such as increasing the "good" HDL cholesterol and reducing the ratio of total cholesterol to HDL cholesterol. The more aerobically fit you are, the higher your HDL cholesterol is likely to be.
- A reduction of blood pressure in people with borderline hypertension. Also, a high level of fitness may protect you from developing high blood pressure at all.
- The more you exercise, the more calories you burn—and the more body fat you lose. Exercise has become a major component in most weight loss programs.
- Higher levels of endurance exercise help control any tendency toward adult-onset diabetes.

- Protection against osteoporosis, or the potentially dangerous decrease in bone mass with age. The heavier your exercise regimen, the less bone mass you tend to lose. Also,
regular exercise encourages the building of new bone.
- An increase in feelings of well-being, including protection
against mental illness, such as depression. Endurance exercise acts as nature's best tranquilizer through the body's
production of endorphins, the morphine-like neurotrans-
mitters that induce feelings of relaxation.

That last benefit—the enhancement of well-being—is probably
the one that is most powerful in making the average exerciser want
more and more of a good thing. But it is not necessary to overdo
your conditioning program, even if your goal is to achieve all of
the above results. You can keep your program in the lower-
intensity range and still receive all the benefits.

How to Achieve All the Benefits of Exercise

Exactly how much exercise is necessary to get all those benefits?
In terms of the fitness points system described in chapter 4, it only
takes thirty-five points a week, using one of the programs described at the end of that chapter. Here is how this level of activity
translates into specific exercise regimens:

- You can run two miles in less than twenty minutes, four
times per week. That equals thirty-six points.
- You can walk three miles in less than forty-five minutes,
five times per week. That translates into forty points.
- You can do aerobic dancing or other continuous exercise
to music for forty-five minutes, four times a week. That's
thirty-six points.

The possibilities are almost limitless. And you do not have to
worry about figuring out the number of fitness points you are
earning, unless you prefer that approach. Just follow the instruc-

tions that place you in the "athletic fitness" category on the charts. Or if you prefer to stay in the lower "health and longevity fitness" category, you will still gain many benefits. With either approach, you will maximize your safety and health by working out on a lower-intensity level.

A personal observation: I have discovered that there is a much greater motivation to continue an exercise program among people who exercise in the "athletic fitness category"—that is, at a level considerably higher than thirty-five points per week. Apparently it takes that amount of energy expenditure to activate the endorphin phenomena. There may even be an addictive effect and some withdrawal symptoms when a rigorous exercise regimen is abruptly discontinued.

Most people, including myself, exercise primarily for the psychological benefits, not for the health and longevity effects. It is those benefits which provide you with the motivation to stay with an exercise program at the higher levels. I will continue to earn fifty to seventy-five points per week simply because that is the level at which I feel the best! That is the level that reduces feelings of depression, increases productivity, and provides an improved self-image. But that level of exercise also requires higher doses of the antioxidant cocktail in order to combat the danger of production of extra free radicals.

APPENDIX

4

The Antioxidants in Your Food— from the Farm to the Store

S ignificant amounts of antioxidants such as vitamin C, beta carotene, and vitamin E may be lost during holding and storage just after they are harvested, or during commercial food processing. On the other hand, taking proper precautions can preserve the antioxidant values in many foods.

As I mentioned in chapter 7, you obviously have less control in preserving the nutritional value of your food before you buy it at your local supermarket than you do after you bring it into your kitchen. Still, if you become aware of how nutrients may be lost in the agricultural and commercial phases of food preparation, you may be able to pick and choose those items that are more likely to contain the maximum amounts of the antioxidants you need.

So here are some of the factors that may affect your food at different stages of storage and processing, before the product ever reaches your kitchen or table. (The information in this appendix is based on a variety of nutritional research sources, including the Cooper Clinic Nutrition Department's files and *Nutritional Evaluation of Food Processing,* third edition, E. Karmas and R. Harris, editors [New York: Van Nostrand Reinhold, 1988]).

The Agricultural Scene: Holding and Storing Foods

Although harvested foods, being biological products, are highly perishable, biological deterioration can be temporarily delayed or

205

blocked by commercial processing. One of the main reasons that food spoils is that "active" water is present in the tissues of many plants and animals. That active type of water can escape easily, in contrast to "structural" water, which remains sealed up in foods. Raw foods with a large amount of active water content, such as leafy vegetables and meats, will deteriorate in just a few days. Raw foods, containing structural water (such as dry seeds), can be stored for years.

After harvesting, the best storage conditions take into account the optimum temperature and humidity for preserving the nutritional values of individual fresh fruits and vegetables. Unfortunately, storage rooms may be used to store more than one commodity, with the result that some foods may not be exposed to the best conditions. Consequently, they may lose much of their nutritional value.

Also, refrigerated trucks usually carry mixed loads, with one temperature for all foods being transported—even though some foods need special, individual storage treatment. Similarly, grocery warehouses and stores tend to keep all their produce at one or perhaps two temperatures, rather than choosing the best temperature for each food.

What are the typical times and temperatures for food storage after harvesting—and what is the impact on their antioxidant values?

Consider citrus fruits. They are generally stored after harvesting for ten to sixteen days at 45 to 48 degrees Fahrenheit. Under those conditions, oranges lose only a slight amount of their vitamin C. Tangerines lose about 25 percent of their vitamin C, but then the losses will cease for as long as eight weeks if they are stored at 32 degrees Fahrenheit.

As for vegetables, vitamin C is easily lost in those that are exposed to air or heat. Rapid cooling methods (vacuum cooling) allow for a longer shelf life for most vegetables and less loss of vitamin C.

The content and maturity of fruits containing vitamin C vary, so it is difficult to make gross generalizations. As a general rule of thumb, however, it is best to eat fruits and vegetables that have not wilted or become noticeably old. Also, some vegetables are more stable in their vitamin C content than others. Sweet peppers, for example, which are available throughout the year, rank fourth in vitamin C content among the forty or more fruits and vegetables that are highest in vitamin C. After distribution and storage under optimum conditions for two to three weeks, the total vitamin C content in sweet peppers varies relatively little. The amount of beta carotene will increase significantly in most fruits as they mature.

What Is the Impact of the Freezing Process?

Frozen fruits and vegetables may actually retain *more* vitamin C than fresh produce because of losses that typically occur to non-frozen items during transportation and handling. Still, there are losses of both vitamin C and beta carotene during commercial processing as a result of two factors: blanching during the freezing process and prolonged storage.

Blanching

Blanching neutralizes enzymes that break down vitamins, including vitamin C and beta carotene. In general, if foods are frozen unblanched, losses during frozen storage will be much greater. The best method of blanching involves steam or microwave techniques, followed by a method of cooling that does not involve water.

Different methods of packaging can also affect vitamin C retention during frozen storage, regardless of the blanching process. In one study involving green beans and carrots stored for twenty-four to forty weeks at 30 degrees Fahrenheit, retention of vitamin C was greater in the unblanched vacuum-packed products than it was in products that were blanched and vacuum-packed. That study indicates that methods of excluding oxygen from frozen products have significant beneficial effects.

Losses of vitamin C may also occur in fruits through the pulling or drawing out of the vitamin into the syrup and during the thawing process. Consequently, it is important to use as much of the syrup or juice in frozen foods as you can.

Storage

It's important to limit the length of time you hold foods after thawing. Thawed raspberries in syrup lose 15 percent of their vitamin C after twenty-four hours at 68 Fahrenheit. Thawed peaches in syrup lose 10 percent of their vitamin C after being kept two hours at room temperature. The most desirable storage temperature for the maximum retention of vitamin C for most frozen vegetables and fruits is 0 degrees Fahrenheit for up to twelve months.

Citrus juice concentrates tend to experience very small losses of vitamin C, probably because of the low pH values (alkalinity) and low oxygen content of these products. Some fruits are fortified with vitamin C prior to freezing so that browning of the food caused by enzymes will be retarded. One study showed that after eight months at 0 degrees Fahrenheit, there was about 80 percent retention of the vitamin C which had been added.

Now, let's sum up. Here is a summary of the impact on antioxidants of each step in the freezing process:

- Prefreezing: With vegetables, 10 to 44 percent of vitamin C may be leached into the water during blanching. With fruits, there will be only slight losses of vitamin C if the food is handled properly.
- Freezing: With both vegetables and fruits, loss of vitamin C is slight.
- Frozen storage (twelve months at 30 degrees Fahrenheit): With vegetables, 10 to 44 percent of vitamin C and 4 to 20 percent of beta carotene may be lost. Losses are greater in

unblanched products. With fruits, 10 to 30 percent of vitamin C may be lost, except in citrus juice concentrate, where losses are minimal.

- Total freezing process (prefreezing, freezing, and storage for six to twelve months at 30 degrees Fahrenheit): With vegetables, there will be an average loss of 45 percent of the original content of vitamin C. Losses of beta carotene may be 4 to 20 percent. With fruits, there should be less than 30 percent loss of vitamin C, with the exception of citrus juice concentrates, which should experience less than 5 percent loss of this nutrient.
- Cooking of frozen food: With vegetables, there would be a 30 percent loss of vitamin C, and a 5 percent loss of beta carotene. The cooking factor would not apply to fruits.

What About Canning?

Losses of vitamin C during the canning of vegetables range from 26 to 75 percent of the original amount of the vitamin. Tomatoes tend to have the lowest percentage of losses, and carrots have the highest losses.

To ensure less than a 10 percent loss of vitamin C and beta carotene, the storage temperature of canned products over a twelve-month period should be 65 degrees Fahrenheit.

If you are alert, you may find that some descriptions of packaging techniques are indicated on food labels or in literature put out by the manufacturer. Here are some points to watch for:

- With milk powders, spray drying produces less loss of vitamin C than does spray rolling. When properly stored, the loss of vitamin C in powdered milk may be only 10 percent after two years.
- Vitamin C and vitamin A, when added to cereals as part of a surface coating on the food, are subject to loss through oxidation (which occurs through exposure to air). But the

209

presence of these two nutrients together enhances their stability. Also, vacuum packaging can help prevent oxidation.

Vitamin C may be lost in the drying process of apple flakes in these amounts: 8 percent during slicing; 62 percent during blanching; and 5 percent during drum drying.

- There is no loss of vitamin C during vacuum drying of tomato concentrate.
- Freeze-drying may produce these beta carotene losses: 13 percent in freeze-dried carrots; and 4 percent in freeze-dried orange juice.
- There are no reports of vitamin E losses during drying.
- Vitamin A is added back to skimmed milk, but alpha-tocopherol (vitamin E) is not. Experimentally, water-soluble forms of vitamin E have been added to skim milk and nonfat dry milk in amounts up to 150 IU per quart. Researchers have found 100 percent stability (no nutritional losses) of the vitamin E for four weeks in the liquid form, and for one year in the dry form.

Summary of Antioxidant Sources, Recommendations, and Effects

Antioxidant	Primary Food Source	Cooper Daily Recommendation*
Vitamin C (ascorbic acid)	Acerola fruit, papaya, oranges, cantaloupe, broccoli, brussels sprouts, grapefruit, strawberries, kiwifruit, cauliflower	500-3,000 mg*
Vitamin E (d-alpha-tocopherol)	Wheat germ, almonds, hazelnuts, mayonnaise, corn oil, cottonseed oil, sunflower oil, egg yolk, butter	200-1,200 IU or mg (1 IU = 1mg)*
Beta Carotene (carotenoid)	Dark green, yellow-orange vegetables and fruits: carrots, sweet potatoes, tomatoes, spinach, squash, cantaloupe, mango, papaya, apricots, broccoli	10,000-50,000 IU (6-30 mg)*
Vitamin A	Milk, eggs, liver, cheese, fish oil, butter	*Not recommended in supplement form* Use only Beta Carotene
Selenium	Seafood, kidney, liver, grains and seeds grown in soil that contains selenium	50-100 mcg* (optional)

Antioxidant	*Primary Food Source*	*Cooper Daily Recommendation**
Coenzyme Q-10	Fish, nuts, lean meats, polyunsaturated fats (also coenzyme Q-10 is manufactured by the body)	*Not recommended*
Probucol (Locelco)	Prescription drug	*Not recommended*

*Refer to p. 127 in text for specific supplement recommendations according to age, sex, and exercise level.

Summary of Antioxidant Sources, Recommendations, and Effects

Antioxidant	RDA	Known and Likely Beneficial Effects
Vitamin C (ascorbic acid)	30 mg (adults) 60 mg (smokers) 45 mg (children)	Promotes wound healing, growth, and tissue repair. Helps in utilization of iron. Combats effects of free radicals. Enhances effect of vitamin E. Reduces risk of certain cancers, especially stomach, but also esophagus, larynx, oral cavity, and pancreas. Lowers risk of cataracts, and may increase immunity to infectious diseases, may lower total cholesterol, and raise the "good" HDL cholesterol.
Vitamin E (d-alpha-tocopherol)	12 IU (women) 15 IU (men) 7 IU (children)	As a strong antioxidant, counters the effects of free radicals. Is an anticoagulant (blood-thinner). Important in formation of blood cells. Helps utilize vitamin K. Reduces risk of certain cancers. May have major benefits in protecting against coronary heart disease and increasing immunity. Reduces risk of cataracts.
Beta Carotene (carotenoid)	No RDA but government assumes that half of vitamin A will come from beta carotene.	Converted into vitamin A in the body (to the extent that vitamin A is needed). It's a carotenoid precursor to vitamin A. As a strong antioxidant, it can lower risk of cataracts, heart disease, and many types of cancer, particularly lung cancer. Also, may protect against bladder cancer, rectal cancer, and melanoma.
Vitamin A	4,000 IU (women) 5,000 IU (men) 2,500 IU (children)	Helps vision; needed in maintaining healthy skin, teeth, bones, and mucous membranes. Helps tissue growth and repair. As an antioxidant, it reduces risk of cancer, particularly lung cancer. May increase resistance to infection.

213

Antioxidant	RDA	Known and Likely Beneficial Effects
Selenium	55 ug (women) 70 ug (men) 20 ug (children)	As a structural element of the enzyme glutathione peroxidase, it protects the intracellular antioxidant glutathione, which works along with vitamins C, E, and beta carotene against peroxidation of cell membranes. May reduce the risk of stomach and esophageal cancer.
Coenzyme Q-10	No RDA	As an antioxidant, may play a role in preventing heart disease and congestive heart failure. Helps recycle vitamin E in the body.
Probucol (Locelco)	No RDA	A strong antioxidant which lowers total and "bad" LDL cholesterol. May protect against cancer, cataracts, and coronary artery disease.

Antioxidant	Possible Side Effects
Vitamin C (ascorbic acid)	At levels greater than 4,000 mg per day, it may cause diarrhea, kidney stones, or liver problems.
Vitamin E (d-alpha-tocopherol)	May increase plasma lipids (fats). At levels above 3,200 IU per day, may cause breast soreness, decreased thyroid hormone levels, diarrhea, headaches, blood pressure elevation, fatigue, blurred vision, or hypoglycemia. Should be taken cautiously with anticoagulants or blood thinners.
Beta Carotene (carotenoid)	Though a non-toxic antioxidant, doses of more than 30 mg (50,000 IU) of beta carotene per day can accumulate under the skin and cause yellow pigmentation. May cause side effects if taken with alcohol. Pregnant women should not take it without physician's approval. Smokers should use caution.
Vitamin A	Doses above 5,000-10,000 IU per day have been associated with toxic liver effects, blurred vision, hair loss, nausea, birth defects, spleen enlargement, and dry skin. Should not be taken at doses above 10,000 IU per day without physician's approval.
Selenium	Garlic odor to breath and sweat. Overdoses associated with hair loss and other toxic effects, such as nausea, vomiting, diarrhea, neuropathy (a disease of the nerve tissue), irritability, or fatigue.
Coenzyme Q-10	No known harmful side effects, but not recommended as a routine antioxidant.
Probucol (Locelco)	Lowers "good" HDL cholesterol. Long-term effects unknown.

Council for Responsible Nutrition: April 13, 1994

1300 19th Street, NW Suite 310
Washington, DC 20036–1609

News Release

THE FINNISH STUDY: PUZZLING RESULTS

A new Finnish study demonstrates that the damage done by a lifetime of smoking cannot be reversed by eleventh hour nutrient intervention. The study on the effect of vitamin E and beta carotene supplements in heavy smokers appears this week in *The New England Journal of Medicine*. In the study, over 29,000 men were given 50 mg of vitamin E, 20 mg of beta carotene, both nutrients, or a placebo for a period of about 6 years. The men had been smoking at least a pack of cigarettes a day for an average of 36 years. Most of them continued to smoke heavily; they were not encouraged to adopt any healthy habits other than supplement use.

"There is no substitute for lifelong healthy habits, including not smoking. Scientists have never suggested that antioxidant nutrients can compensate for high-risk behaviors," said Annette Dickinson, Ph.D., CRN Director of Scientific and Regulatory Affairs.

The study found no effect of vitamin E supplementation on

lung cancer in these men, and found a possible negative effect of beta carotene supplementation. The 14,000 men who took beta carotene had 72 more cases of lung cancer than the 14,000 men who did not (474 cases vs 402 cases). Could this difference have occurred by chance?

The authors say that chance is a possible explanation: "We are aware of no other data at this time . . . that suggest harmful effects of beta carotene, whereas there are data indicating benefit. Furthermore, there are no known or described mechanisms of toxic effects of beta carotene, no data from animal studies suggesting beta carotene toxicity, and no evidence of serious toxic effects of this substance in humans. In light of all the data available, an adverse effect of beta carotene seems unlikely; in spite of its formal statistical significance, therefore, this finding may well be due to chance."

The authors provide data on other cancers and other disease conditions, but do not indicate whether any differences that exist are significant. For example, the 14,000 men who took vitamin E had 52 fewer cases of prostrate cancer and 13 fewer cases of colorectal cancer than the men who were not given vitamin E. They had 22 more cases of hemorrhagic stroke but 11 fewer cases of ischemic stroke. Are these differences just the luck of the draw— the expected variation among subgroups when one studies 29,000 men over a period of 6 years—or do they mean something? Unfortunately, we cannot know the answer, based on the information provided in the article.

Totality of the available data

The Finnish study must be evaluated in the context of all the available data. There is an abundance of evidence demonstrating that people who have generous intakes of several antioxidant nutrients—vitamin E, carotenoid, and vitamin C—have a decreased risk of numerous chronic diseases, including cancers of many types, cardiovascular disease, and cataracts. Optimum antioxidant intake could potentially not only save lives and reduce the risk of debilitating diseases, but could also have a substantial impact on health care costs. A recent economic analysis commissioned by CRN suggested that annual hospitalization savings in the U.S. alone could amount to almost $9 billion.

Many physicians, nutritionists, and public health experts believe it makes sense, based on the totality of the available data, to increase dietary intakes of these nutrients, and to use supplements of these nutrients in a rational manner.

No evidence of adverse effects of antioxidants in U.S. studies

Major clinical and epidemiological studies in the U.S., involving more than 140,000 subjects, have revealed no adverse effects of the antioxidant nutrients. There has been no suggestion of problems with the Physicians' Health Trial, in which more than 20,000 U.S. physicians have taken 50 mg of beta carotene every other day for over 10 years. In fact, in a subgroup of those physicians who had heart disease, beta carotene was found to reduce the risk of coronary events by more than 40 percent. Cancer results from this trial have not yet been published. In the physicians' trial as in all intervention trials, health effects as well as potential adverse effects are regularly and routinely monitored. In epidemiological studies from Harvard involving more than 80,000 nurses and about 40,000 male health professionals, vitamin E supplement use was associated with a 40 percent decreased risk of heart disease, and generous intakes of beta carotene were also significantly protective, especially in smokers. In several intervention trials, beta carotene has been shown to help prevent and even to treat precancerous lesions of the oral cavity.

Vitamin E levels may have been too low

The Finnish study used only modest levels of vitamin E (50 mg per day). In contrast, other studies being supported by the National Cancer Institute are using vitamin E levels up to 12 times higher. Moreover, the two above-mentioned epidemiological studies from Harvard showed a protective effect of vitamin E supplements against heart disease risk only in persons who took at least 100 IU per day for at least 2 years.

The prevention paradigm vs drug treatment

The new paradigm in health promotion is to prevent chronic disease through lifetime healthy habits, including improved nutrition. It is a matter of some concern that this paradigm is being tested in numerous intervention trials based on a drug treatment model. Cancer has a long latency period, and it has been estimated that the interval between the initiation of lung cancer and its eventual diagnosis is in the range of 8 to 20 years. Trials of supplementation with individual nutrients in high-risk populations late in the disease process may not be true tests of the preventive potential of lifelong generous intakes of those nutrients, in conjunction with other healthy habits. This issue needs further consideration.

Contact CRN for additional information

For more information on studies relating to antioxidant nutrients and for the names and phone numbers of key antioxidant researchers who are familiar with the data, contact the Council for Responsible Nutrition (CRN). CRN is a trade association of nutritional supplement manufacturers.

References

The following references are grouped according to the chapters of the book for which they have served as source materials. Also, there are indications in the appendices as to which chapter references apply to a specific appendix. A brief citation may be given in the main text for a finding, report, or quotation in the text, and then readers who want a more complete reference will often find what they need in this section.

Chapter One: The Antioxidant Revolution

American Journal of Clinical Nutrition, Supplement to Vol. 53, No. 1, Jan. 1991, pp. 189S-396S. The following articles in this supplement have provided background for the entire text. The articles are listed according to their order or appearance in the supplement.

1. Free radical formation and tissue damage: antioxidant defense systems.

   Diplock, A.T., "Antioxidant nutrients and disease prevention: an overview."

   Di Mascio, P., Murphy, M.E., and Sies, H., "Antioxidant defense systems: the role of carotenoids, tocopherols, and thiols."

   Niki, E., Yamamoto, Y., Komuro, E., and Sato, K., "Membrane damage due to lipid oxidation."

   Luc, G., and Fruchart, J.C., "Oxidation of lipoproteins and atherosclerosis."

   Yoshikawa, T., Yasuda, M., Ueda, S., Naito, Y., Tanigawa, T., Oyamada, H., and Kondo, M., "Vitamin E in gastric mucosal injury induced by ischemia reperfusion."

   Ferrari, R., Ceconi, C., Curello, S., Cargnoni, A., Pasini, E., De Guili, F., and Altertini, A., "Role of oxygen free radicals in ischaemic and reperfused myocardium."

2. The antioxidant vitamins and beta carotene in cancer prevention.

   Tubiana, M., "Human carcinogenesis—introductory remarks."

   Weisburger, J.H., "Nutritional approach to cancer prevention with emphasis on vitamins, antioxidants, and carotenoids."

   Krinsky, N.I., "Effects of carotenoids in cellular and animal systems."

   Tannenbaum, S.R., Wishnok, J.S., and Leaf, C.D., "Inhibition of nitrisamine formation by ascorbic acid."

   Ziegler, R.G., "Vegetables, fruits, and carotenoids and the risk of cancer."

   Comstock, G.W., Helzisouer, K.J., and Bush, T.L., "Prediagnostic serum levels of carotenoids and vitamin E as related to subsequent cancer in Washington County, Maryland."

   Stahelin, H.B., Gey, K.F., Eichholzer, M., and Ludin, E., "Beta-carotene and cancer prevention: the Basel Study."

   Block, G., "Vitamin C and cancer prevention: the epidemiologic evidence."

   Knekt, P., Aromaa, A., Maatela, J., Aaran, R.K., Nikkari, T., Hakama, M., Hakulinen, T., Peto, R., and Teppo, L., "Vitamin E and cancer prevention."

   Schorah, C.J., Sobala, G.M., Sanderson, M., Collis, N., and Primrose, J.N., "Gastric juice ascorbic acid: effects of disease and implications for gastric carcinogenesis."

   Garewal, H.S., "Potential role of beta-carotene in prevention of oral cancer."

   Stich, H.F., Matthew, B., Sankaranarayanaaan, R., and Krishnan Nair, M., "Remission of precancerous lesions in the oral cavity of tobacco chewers and maintenance of the protective effect of beta-carotene or vitamin A."

   Malone, W.F., "Studies evaluating antioxidants and beta-carotene as chemopreventives."

3. The prevention of cardiovascular disease.

   Esterbauer, H., Dieber-Rotheneder, M., Striegl, G., and

Waeg, G., "Role of vitamin E in preventing the oxidation of low-density lipoprotein."

Trout, D.L., "Vitamin C and cardiovascular risk factors."

Gey, K.F., Puska, P., Jordan, P., and Moser, U.K., "Inverse correlation between plasma vitamin E and morality from ischemic heart disease in cross-cultural epidemiology."

4. The prevention of cataract formation.

Varma, S.D., "Scientific basis for medical therapy of cataracts by antioxidants."

Robertson, J. McD., Donner, A.P., and Trevithick, J.R., "A possible role for vitamins C and E in cataract prevention."

Jacques, P.F., and Chylack, L.T. Jr., "Epidemiologic evidence of a role for the antioxidant vitamins and carotenoids in cataract prevention."

5. Antioxidant vitamins and beta carotene: their roles in the future.

Block, G., "Dietary guidelines and the results of food consumption surveys."

Anderson, R., "Assessment of the roles of vitamin C, vitamin E, and beta-carotene in the modulation of oxidant stress mediated by cigarette smoke-activated phagocytes."

Merry, P., Grootveld, M., Lunec, J., and Blake, D.R., "Oxidative damage to lipids within the inflamed human joint provides evidence of radical-mediated hypoxic reperfusion injury."

Chiswick, M., Gladman, G., Sinha, S., Toner, N., and Davies, J., "Vitamin E supplementation and periventricular hemorrhage in the newborn."

Cutler, R.G., "Antioxidants and aging."

Fahn, S., "An open trial of high-dosage antioxidants in early Parkinson's disease."

Schmidt, K., "Antioxidant vitamins and beta-carotene: effects on immunocompetence."

Singh, V.N., and Gaby, S.K., "Premalignant lesions: role of antioxidant vitamins and beta-carotene in risk reduction and prevention of malignant transformation."

Pryor, W.A., "The antioxidant nutrients and disease prevention—what do we know and what do we need to find out?"

Slater, T.F., "Concluding remarks."

Angier, Natalie, "Free radicals: the price we pay for breathing," *The New York Times Magazine*, April 25, 1993, p. 62ff.

Bast, et al., "Oxidants and antioxidants: state of the art," *The American Journal of Medicine*, Sept. 30, 1991, Vol. 91, Supple. 3C, p. 3C-2Sff.

Crystal, Ronald G., "Introduction," *The American Journal of Medicine*, Sept. 30, 1991, Vol. 91, Supple. 3C, p. 3C-1S.

Crystal, Ronald G., "Summary," *The American Journal of Medicine*, Sept. 30, 1991, Vol. 91, Supple. 3C, p. 3C-145S.

Jenkins, R.R., et al., "Influence of exercise on clearance of oxidant stress products and loosely bound iron," *Medicine and Science in Sports and Exercise*, 1993, Vol. 25, No. 2, pp. 213-217.

Packer, Lester, "Protective role of vitamin E in biological systems," *American Journal of Clinical Nutrition*, 1991, Vol. 53, pp. 1050S-5S.

Scandalios, John G., "Molecular biology of free radical scavenging systems," *Free Radical Biology & Medicine*, 1993, Vol. 14, p. 227.

Sies, Helmut, "Antioxidant functions of vitamins—vitamins E and C, beta-carotene, and other carotenoids," *Nutrition Today*, July/August 1990, p. 7ff.

Sies, Helmut, "Oxidative Stress: from basic research to clinical application," *The Amercian Journal of Medicine*, Sept. 30, 1991, Vol. 91, Supple. 3C, p. 3C-31Sff.

Stipp, David, "Heart-attack study adds to the cautions about iron in the diet," *The Wall Street Journal*, Sept. 8, 1992, p. 1.

"Vitamin E for a Healthy Heart," *Newsweek*, May 31, 1993, p. 62.

Ward, Peter A., "Mechanisms of endothelial cell killing by H202 or products of activated neutrophils," *The American Journal of Medicine*, Sept. 30, 1991, Vol. 91, Supple. 3C., p. 3C-89Sff.

Chapter Two: Unmasking the Free Radical Threat: The Latest News From the Medical Front

Abrams, Jonathan, "Interactions between organic nitrates and thiol group," *The American Journal of Medicine*, Sept. 30, 1991, Vol. 91, Supple. 3C, p. 3C-106Sff.

Ambrosio, Guiseppe, et al., "Myocardial reperfusion injury: mechanisms and management—a review," *The American Journal of Medicine*, Sept. 30, 1991, Vol. 91, Supple. 3C, p. 3C-86Sff.

American Journal of Clinical Nutrition, Supplement to Vol. 53, No. 1, Jan. 1991, pp. 189S-396S. The following articles in this supplement have provided background for the text. The articles are listed according to their order of appearance in the supplement.

1. Free radical formation and tissue damage: antioxidant defense systems.

Diplock, A.T., "Antioxidant nutrients and disease prevention: an overview."

Di Mascio, P., Murphy, M.E., and Sies, H., "Antioxidant defense systems: the role of carotenoids, tocopherols, and thiols."

Niki, E., Yamamoto, Y., Komuro, E., and Sato, K., "Membrane damage due to lipid oxidation."

Luc, G., and Fruchart, J.C., "Oxidation of lipoproteins and atherosclerosis."

Yoshikawa, T., Yasuda, M., Ueda, S., Naito, Y., Tanigawa T., Oyamada, H., and Kondo, M., "Vitamin E in gastric mucosal injury induced by ischemia-reperfusion."

Ferrari, R., Ceconi, C., Curello, S., Cargnoni, A., Pasini, E., De Guili, F., and Altertini, A., "Role of oxygen free radicals in ischemic and reperfused myocardium."

2. The antioxidant vitamins and beta carotene in cancer prevention.

Tubiana, M., "Human carcinogenesis—introductory remarks."

Weisburger, J.H., "Nutritional approach to cancer prevention with emphasis on vitamins, antioxidants and carotenoids."

Krinsky, N.I., "Effects of carotenoids in cellular and animal systems."

Tannenbaum, S.R., Wishnok, J.S., and Leaf, C.D., "Inhibition of nitrisamine formation by ascorbic acid."

Ziegler, R.G., "Vegetables, fruits, and carotenoids and the risk of cancer."

Comstock, G.W., Helzisouer, K.J., and Bush, T.L., "Prediagnostic serum levels of carotenoids and vitamin E as related to subsequent cancer in Washington County, Maryland."

Stahelin, H.B., Gey, K.F., Eichholzer, M., and Ludin, E., "Beta-carotene and cancer prevention: the Basel Study."

Block, G., "Vitamin C and cancer prevention: the epidemiologic evidence."

Knekt, P., Aromaa, A., Maatela, J., Aaran, R.K., Nikkari, T., Hakama, M., Hakulinen, T., Peto, R., and Teppo, L., "Vitamin E and cancer prevention."

Schorah, C.J., Sobala, G.M., Sanderson, M., Collis, N., and Primrose, J.N., "Gastric juice ascorbic acid: effects of disease and implications for gastric carcinogenesis."

Garewal, H.S., "Potential role of beta-carotene in prevention of oral cancer."

Stich, H.F., Matthew, B., Sankaranarayanaaan, R., and Krishnan Nair, M., "Remission of precancerous lesions in the oral cavity of tobacco chewers and maintenance of the protective effect of beta-carotene or vitamin A."

Malone, W.F., "Studies evaluating antioxidants and beta-carotene as chemopreventives."

3. The prevention of cardiovascular disease.

Esterbauer, H., Dieber-Rotheneder, M., Striegl, G., and Waeg, G., "Role of vitamin E in preventing the oxidation of low-density lipoprotein."

Trout, D.L., "Vitamin C and cardiovascular risk factors."

Gey, K.F., Puska, P., Jordan, P., and Moser, U.K., "Inverse correlation between plasma vitamin E and mortality from ischemic heart disease in cross-cultural epidemiology."

4. The prevention of cataract formation.

Varma, S.D., "Scientific basis for medical therapy of cataracts by antioxidants."

Robertson, J. McD., Donner, A.P., and Trevithick, J.R., "A possible role for vitamins C and E in cataract prevention."

Jacques, P.F., and Chylack, L.T. Jr., "Epidemiologic evidence of a role for the antioxidant vitamins and carotenoids in cataract prevention."

5. Antioxidant vitamins and beta-carotene: their roles in the future.

Block, G., "Dietary guidelines and the results of food consumption surveys."

Anderson, R., "Assessment of the roles of vitamin C, vitamin E, and beta-carotene in the modulation of oxidant stress mediated by cigarette smoke-activated phagocytes."

Merry, P., Grootveld, M., Lunec, J., and Blake, D.R., "Oxidative damage to lipids within the inflamed human joint provides evidence of radical-mediated hypoxic reperfusion injury."

Chiswick, M., Gladman, G., Sinha, S., Toner, N., and Davies, J., "Vitamin E supplementation and periventricular hemorrhage in the newborn."

Cutler, R.G., "Antioxidants and aging."

Fahn, S., "An open trial of high-dosage antioxidants in early Parkinson's disease."

Schmidt, K., "Antioxidant vitamins and beta-carotene: effects on immunocompetence."

Singh, V.N., and Gaby, S.K., "Premalignant lesions: role of antioxidant vitamins and beta-carotene in risk reduction and prevention of malignant transformation."

Pryor, W.A., "The antioxidant nutrients and disease prevention—what do we know and what do we need to find out?"

Slater, T.F., "Concluding remarks."

Angier, Natalie, "Free radicals: the price we pay for breathing," *The New York Times Magazine*, April 25, 1993, p. 62ff.

Bolton-Smith, C., et al., "The Scottish heart health study. Dietary intake by food frequency questionnaire and odds ratios for coronary heart disease risk." II. "The antioxidant vitamins and fibre," *European Journal of Clinical Nutrition*, 1992, Vol. 46, pp. 85–93.

Byers, Tim, and Perry, Geraldine, "Dietary carotenes, vitamin C, and vitamin E as protective antioxidants in human cancers," *Annual Review of Nutrition*, 1992, Vol. 12, pp. 139–59.

Byers, et al., "New directions: the diet-cancer link," *Patient Care*, Nov. 30, 1990, p. 34ff.

Cacciuttolo, Marco A., et al., "Hyperoxia induces DNA damage in mammalian cells," *Free Radical Biology & Medicine*, 1993, Vol. 14, pp. 267–276.

Canfield, Louise M., et al., "Carotenoids as cellular antioxidants," *Proceedings of the Society for Experimental Biology and Medicine*, 1992, Vol. 200, p. 260ff.

Chen, J., et al., "Antioxidant status and cancer mortality in China," *International Journal of Epidemiology*, 1992, Vol, 21, No. 4, p. 625ff.

Cochrane, Charles G., "Cellular injury by oxidants," *The American Journal of Medicine*, Sept. 30, 1991, Vol. 91, Supple. 3C, p. 3C-23Sff.

Comstock, George W., et al., "Serum retinol, beta-carotene, vitamin E, and selenium as related to subsequent cancer of specific sites," *American Journal of Epidemiology*, 1992, Vol. 135, No. 2, p. 115ff.

"The data support a role for antioxidants in reducing cancer risk," *Nutrition Reviews*, July 1992, Vol. 50, No. 7, p. 207ff.

Edgington, Stephen M., "Chemokines in cardiovascular disease," *Bio/Technology*, June 1993, Vol. 11, p. 676ff.

Esterbauer, Hermann, et al., "Effect of antioxidants on oxidative modification of LDL," *Annals of Medicine*, 1991, Vol. 23, p. 573–581.

Esterbauer, Hermann, et al., "The role of lipid peroxidation and antioxidants in oxidative modification of LDL," *Free Radical Biology & Medicine*, 1992, Vol. 13, pp. 341–390.

Ferrari, Roberto, et al., "Oxygen free radicals and myocardial damage: protective role of thiol-containing agents," *The American Journal of Medicine*, Sept. 30, 1991, Vol. 91, Supple. 3C, p. 3C-95Sff.

Flaherty, Johyn T., "Myocardial injury mediated by oxygen free radicals," *The American Journal of Medicine*, Sept. 30, 1991, Vol. 91, Supple. 3C., p. 3C-79Sff.

Flora, Silvio De, et al., "Antioxidant activity and other mechanisms of thiols involved in chemoprevention of mutation and cancer," *The American Journal of Medicine*, Sept. 30, 1991, Vol. 91, Supple. 3C, p. 3C-122Sff.

Gaziano, J. Michael, et al., "Dietary antioxidants and cardiovascular disease," *Annals New York Academy of Sciences*, 1992, Vol. 669, p. 249ff.

Halliwell, Barry, and Gutteridge, John M.C., *Free Radicals in Biology and Medicine*. Oxford: Clarendon Press, 1985, pp. 1–66.

Halliwell, Barry, "Reactive oxygen species in living systems: source, biochemistry, and role in human disease," *The American Journal of Medicine*, Sept. 30, 1991, Vol. 91, Supple. 3C, p. 3LC-14Sff.

Harris, William S., "The prevention of atherosclerosis with antioxidants," *Clinical Cardiology*, 1992, Vol. 15, pp. 636–640.

Hearse, David J., et al., "Prospects for antioxidant therapy in cardiovascular medicine," *The American Journal of Medicine*, Sept. 30, 1991, Vol. 91, Supple. 3C, p. 3C-118Sff.

Horowitz, John D., "Thiol-containing agents in the management of unstable angina pectoris and acute myocardial infarction," *The American Journal of Medicine*, Sept. 30, 1991, Vol. 91, p. 3C-113Sff.

Jaakkola, K., et al., "Treatment with antioxidant and other nutrients in combination with chemotherapy and irradiation in patients with small-cell lung cancer," *Anticancer Research*, 1992, Vol. 12, pp. 599–606.

Jenkins, R.R., et al., "Introduction: oxidant stress, aging, and exercise," *Medicine and Science in Sports and Exercise: Official Journal of the American College of Sports Medicine*, 1993, Vol. 25, No. 2, pp. 210–212.

Jialal, I., and Grundy, S.M., "Influence of antioxidant vitamins of LDL oxidation," *Annals New York Academy of Sciences*, p. 237.

Kanter, M.M., et al., "Effects of exercise training on antioxidant enzymes and cariotoxicity of doxorubicin," *Journal of Applied Physiology*, 1985, Vol. 59(4), pp. 1298–1303.

Krinsky, Norman I., "Mechanism of action of biological antioxidants," *Proceedings of the Society for Experimental Biology and Medicine*, 1992, Vol. 200, p. 248ff.

Laranjinha, Joao A.N., et al., "Lipid peroxidation and its inhibition in low density lipoproteins: quenching of cis-parinaric acid fluorescence," *Archives of Biochemistry and Biophysics*, August 15, 1992, Vol. 297, No. 1, pp. 147–154.

Leary, Warren E., "Vitamins cut cancer deaths in large study in rural China," *The New York Times*, Sept. 15, 1993, p. C13.

Liebman, Bonnie, "Antioxidants and cancer," *Nutrition Action Healthletter*, July/August 1992, p. 1ff.

MacNee, William, et al., "The effects of N-acetylcysteine and glutathione on smoke-induced changes in lung phagocytes and epithelial cells," *The American Journal of Medicine*, Sept. 30, 1991, Vol. 91, Supple. 3C, p. 3C-6Sff.

Martins, Elizabeth A. L., "Role of antioxidants in protecting cellular DNA from damage by oxidative stress," *Mutation Research*, 1991, Vol. 250, pp. 95–101.

Mino, Makoto, "Clinical uses and abuses of vitamin E in children," *Proceedings of the Society for Experimental Biology and Medicine*, 1992, Vol. 200, p. 266ff.

Prasad, Kailash, et al., "Oxygen free radicals and hypercholesterolemic atherosclerosis: effect of vitamin E," *American Heart Journal*, 1993, Vol, 125, p. 958ff.

Riemersma, Rudolph A., et al., "Plasma antioxidants and coronary heart disease: vitamins C and E, and selenium," *European Journal of Clinical Nutrition*, 1990, Vol. 44, pp. 143–150.

Rimm, Eric B., "Vitamin E consumption and the risk of coronary heart disease in men," *The New England Journal of Medicine*, May 20, 1993, Vol. 328, pp. 1450–6.

Stampfer, Meir J., et al., "Vitamin E consumption and the risk of coronary disease in women," *The New England Journal of Medicine*, May 20, 1993, Vol. 328, pp. 1444–9.

Steinberg, Daniel, et al., "Antioxidants in the prevention of human atherosclerosis," Summary of the Proceedings of a National Heart, Lung, and Blood Institute Workshop: September 5–6, 1991, Bethesda, Maryland.

Stipp, David, "Studies showing benefits of antioxidants prove potent tonic for sales of vitamin E," *The Wall Street Journal*, April 13, 1993, p. B1.

"A symposium: antioxidants and lipid metabolism," *The Ameri-*

can Journal of Cardiology, February 25, 1993, Gotto, Antonio M., editor.

The following presentations are included:

"Introduction" by Antonio M. Gotto, Jr.

"Overview of current issues in management of dyslipidemia," by Antonio M. Gotto, Jr.

"A modern view of atherogenesis," by Colin J. Schwartz, et al.

"The role of lipids and antioxidative factors for development of atherosclerosis," by Goran Walldius, et al.

"Femoral and coronary angiographic trials," by Linda Cashin-Hemphill.

"Glycation and oxidation: a role in the pathogenesis of atherosclerosis," by Timothy J. Lyons

"Implications for the present and direction for the future," by Michael H. Davidson.

"Discussion."

"Vitamins C & E wage war against atherosclerosis," *Nutrition & Health News,* Center for Human Nutrition, The University of Texas Southwestern Medical Center at Dallas, Fall 1990, Vol. VII, No. 3, p. 1.

Waldholz, Michael, "Evidence grows that nutrients prevent cancer," *The Wall Street Journal,* Sept. 15, 1993, p. B1.

Yoshida, Lucia Satiky, et al., "Phosphatidylcholine peroxidation and liver cancer in mice fed a choline-deficient diet with ethionine," *Free Radical Biology & Medicine,* 1993, Vol. 14, pp. 191–199.

Chapter Three: Designing Your Personal Defense Against the Molecular Outlaws

Adams, James D., Jr., et al., "Oxygen free radicals and Parkinson's disease," *Free Radical Biology & Medicine,* 1991, Vol. 10, pp. 161–169.

Allegra, Luigi, et al., "Ozone-induced impairment of mucociliary transport and its prevention for N-Acetylcysteine," *The American Journal of Medicine,* Sept. 30, 1991, Vol. 91, Supple. 3C, p. 3C-67Sff.

"Antioxidants and HIV infection," *Nutrition Reviews,* June 1992, Vol. 50, No. 6, p. 180.

"Antioxidants: What are they? Can they help keep you healthy?" *Mayo Clinic Health Letter*, Vol. 11, No. 8, August 1993, p. 1ff.

Crystal, Ronald G., "Oxidants and respiratory tract epithelial injury: pathogenesis and strategies for therapeutic intervention," *The American Journal of Medicine*, Sept. 30, 1991, Vol. 91, Supple. 3C, p. 3C-39Sff.

Droge, Wulf, "Modulation of lymphocyte functions and immune responses by cysteine and cysteine derivatives," *The American Journal of Medicine*, Sept. 30, 1991, Vol. 91, Supple 3C, p. 3C-140Sff.

"Finding may show the path of AIDS," *The New York Times*, Nov. 14, 1993, p. 27A.

Flanagan, Robert J., "Use of Ne-acetylcysteine in clinical toxocology," *The American Journal of Medicine*, Sept. 30, 1991, Vol. 91, Supple. 3C, p. 3C-131Sff.

Shoulson, Ira, "Antioxidative therapeutic strategies for Parkinson's disease," *Annals New York Academy of Sciences*, p. 37ff.

Taylor, Allen, "Role of nutrients in delaying cataracts," *Annals of New York Academy of Sciences*, p. 111ff.

Varma, Shambhu D., "Scientific basis for medical therapy of cataracts by antioxidants," *American Journal of Clinical Nutrition*, 1991, Vol. 23, pp. 335S-45S.

Chapter Four: The Lower-intensity Exercise Program

Alessio, Helaine M., "Exercise-induced oxidative stress," *Medicine and Science in Sports and Exercise*, 1993, Vol. 25, No. 2, pp. 218–224.

Alessio, Helaine M., et al., "Lipid peroxidation and scavenger enzymes during exercise: adaptive response to training," *Journal of Applied Physiology*, 1988, Vol. 64, No. 4, pp. 1333–1336.

Cao, Guohua, et al., "Oxygen-radical absorbance capacity assays for antioxidants," *Free Radical Biology & Medicine*, 1993, Vol, 14, pp. 303–311.

Dernbach, A.R., et al., "No evidence of oxidation stress during high-intensity rowing training," *Journal of Applied Physiology*, 1993, Vol. 74, No. 5, pp. 2140–2145.

Dillard, C.J., et al., "Effects of exercise, vitamin E, and ozone on pulmonary function and lipid peroxidation," *Journal of Applied Physiology*, 1978, Vol. 45, No. 6, pp. 927–932.

"Experts release new recommendation to fight America's epidemic of physical inactivity," *News Release*, American College of Sports Medicine, July 29, 1993.

Gohil, Kishorchandra, et al., "Effect of exercise training on tissue vitamin E, and ubiquinone content," *Journal of Applied Physiology*, 1987, Vol. 63, No. 4, pp. 1638–1641.

Gohil, K., et al., "Vitamin E deficiency and vitamin C supplements: exercise and mitochondrial oxidation," *Journal of Applied Physiology*, 1986, Vol. 60, No. 6, pp. 1986–1991.

Ji, Li Li, et al., "Blood glutathione status during exercise: effect of carbohydrate supplementation," *Journal of Applied Physiology*, 1993, Vol. 74, No. 2, pp. 788–792.

Ji, Li Li, et al., "Responses of glutathione system and antioxidant enzymes to exhaustive exercise and hydroperoxide," *Journal of Applied Physiology*, 1992, Vol. 72, No. 2, pp. 549–554.

Kihlstrom, M.T., "Lipid peroxidation capacities in the myocardium of endurance-trained rats and mice in vitro," *Acta Physiol Scand*, 1992, Vol. 146, pp. 177–183.

Lovlin, R., et al., "Are incidents of free radical damage related to exercise intensity?" *European Journal of Applied Physiology*, 1987, Vol. 56, pp. 313–316.

Maughan, Ronald J., et al., "Delayed onset muscle damage and lipid peroxidation in man after a downhill run," *Muscle & Nerve*, April 1989, Vol. 12, pp. 332–336.

McKinsey, David S., "Privileged information," *Bottom Line Personal*, August 15, 1993, p. 9ff.

Ohno, Hideki, et al., "Effect of brief physical exercise on the concentrations of immunoreactive superoxide dismutase isoenzymes in human plasma," *Tohoku Journal of Experimental Medicine*, 1992, Vol. 167, pp. 301–303.

Quiroga, Gustavo Barja de, "Brown fat thermogenesis and exercise: two examples of physiological oxidative stress?" *Free Radical Biology & Medicine*, 1992, Vol. 13, pp. 325–340.

Sjodin, Bertil, et al., "Biochemical mechanisms for oxygen free radical formation during exercise," *Sports Medicine*, 1990, Vol. 10, No. 4, pp. 236–254.

Singh, Vishwa N., "A current perspective on nutrition and exercise," *Journal of Nutrition*, 1992, Vol. 122, pp. 760–765.

Chapter Five: The Strength Training Program

"Antioxidants and the elite athlete," proceedings of panel discussion during the 39th annual meeting of the American College of Sports Medicine, May 27, 1992.

Goldfarb, Allan H., "Antioxidants: role of supplementation to prevent exercise-induced oxidative stress," *Medicine and Science in Sports and Exercise*, 1993, Vol. 25, No. 2, pp. 232–236.

Gordon, Neil F., *Chronic Fatigue: Your complete exercise guide*. Dallas: Human Kinetics Publishers, 1993.

Jenkins, R.R., et al., "Introduction: oxidant stress, aging, and exercise," *Medicine and Science in Sports and Exercise*, 1993, Vol. 25, No. 2, pp. 210–212.

Noakes, Tim, *Lore of Running*, third edition. Champaign, Ill: Leisure Press, 1991.

Packer, Lester, "Protective role of vitamin E in biological systems," *American Journal of Clinical Nutrition*, 1991, Vol. 53, pp. 1050S-5S.

Singh, Vishwa N., "A current perspective on nutrition and exercise," *Journal of Nutrition*, 1992, Vol. 122, pp. 760–765.

Sjodin, Bertil, "Biochemical mechanisms for oxygen free radical formation during exercise," *Sports Medicine*, 1990, Vol. 10, No. 4, pp. 236–254.

The Strength Connection. Dallas: Cooper Institute for Aerobics Research, 1990.

"Vitamin as muscle-damage fighter," *The New York Times*, Oct. 31, 1992. p. C12.

Chapter Six: The Antioxidant Cocktail—Advantages, Side Effects, and Variations

"Alcohol and beta-carotene: a cocktail lethal to the liver," *Environmental Nutrition*, February 1993, p. 8.

Armstrong, Francie, "Nutrient primer: vitamin E," *Runner's World*, April, 1993, p. 19.

Buettner, Garry R., et al., "Ascorbate free radical as a marker of oxidative stress: an EPR study," *Free Radical Biology & Medicine*, 1993, Vol. 14, pp. 49–55.

Burros, Marian, "Take your vitamins and eat your veggies," *Health Confidential*, July 1993, pp. 3–4.

"Dietary versus cellular zinc: the antioxidant paradox," Letters to the editor, *Free Radical Biology & Medicine*, 1993, Vol. 14, pp. 95–97.

"Don't be fooled! Know the difference between natural and synthetic vitamin E," J.R. Carlson Laboratories, Inc., 1991.

"Heavy metal and the heart," *Harvard Health Letter*, Dec. 1992, Vol. 18, No. 2.

Hennekens, Charles H., "Antioxidants: do they decrease the risk of cardiovascular disease?" *Nutrition & the M.D.*, August 1993, Vol. 19, No. 8, pp. 1–2.

Levander, Orville A., "Selenium and sulfur in antioxidant protective systems: relationships with vitamin E and malaria," *Proceedings of the Society for Experimental Biology and Medicine*, 1992, Vol. 200, p. 255ff.

Olson, James A., et al., "Antioxidants in health and disease: overview," *Proceedings of the Society for Experimental Biology and Medicine*, 1992, Vol. 200, p. 245ff.

Packer, Lester, "Interactions among antioxidants in health and disease: vitamin E and its redox cycle," *Proceedings of the Society for Experimental Biology and Medicine*, 1992, Vol. 200, p. 271ff.

Packer, Lester, "Protective role of vitamin E in biological systems," *American Journal of Clinical Nutrition*, 1991, Vol. 53, pp. 1050S-5S.

Palozza, Paola, et al., "Communication: beta carotene and alpha tocopherol are synergistic antioxidants," *Archives of Biochemistry and Biophysics*, August 15, 1992, Vol. 297, No. 11, pp. 184–187.

Rice-Evans, Catherine A., et al., "Current status of antioxidant therapy," *Free Radical Biology & Medicine*, 1993, Vol. 15, pp. 77–96.

Smith, Trevor, "Eat for life," *Running & FitNews*, Nov. 1991, Vol. 9, No. 11, p. 4.

Sukalski, Katherine A., et al., "Decreased susceptibility of liver mitochondria from diabetic rats to oxidative damage and associated increase in alpha-tocopherol," *Free Radical Biology & Medicine*, 1993, Vol. 14, pp. 57–65.

"Supplement-taker's guide to the universe," *Nutrition Action Health Letter*, January/February 1993, Vol. 10, No. 1. (entire issue)

"Vitamins C & E wage war against atherosclerosis," *Nutrition & Health News*, Center for Human Nutrition, The University of Texas Southwestern Medical Center at Dallas, Fall 1990, Vol. VII, No. 3, p. 1ff.

Chapter Seven: Cooking and Eating the Antioxidant Way

"Antioxidants in a cup of tea?" *University of California at Berkeley Wellness Letter*, January 1992, p. 1.

Applegate, Liz, "Supplement Speak," *Runner's World*, March 1993, pp. 22–24.

Blumberg, Jeffrey G., "Dietary antioxidents and aging," *Contemporary Nutrition*, 1992, Vol. 17, No. 3, p. 1.

"Can taking supplements help you ward off disease?" *Tufts University Diet and Nutrition Letter*, April 1991, Vol. 9, No. 2, p. 3.

"Cancer protection in our food: antioxidants fight cancer-causing cell damage," *American Institute for Cancer Research Newsletter*, Spring 1993, Issue 39, p. 4.

Carlson, Beth L., et al., "Loss of vitamin C in vegetables during the foodservice cycle," *Journal of American Dietetic Association*, January 1988, Vol. 88, p. 65–67.

Dietz, Jane M., et al., "Effects of thermal processing upon vitamins and proteins in foods," *Nutrition Today*, July/August, 1989, p. 6–14.

Erdman, John W., Jr., et al., "Factors affecting the bioavailability of vitamin A, carotenoids, and vitamin E," *Food Technology*, October 1988, p. 214ff.

Foods & Nutrition Encyclopedia, first edition. Clovis, California: Pegus Press, 1983. Vol. 2, pp. 2232–2269.

"Free radicals and antioxidants: finding the key to heart disease, cancer, and the aging process," *Wellness Letter,* October 1991, Vol. 18, Issue 1, pp. 4–5.

"Green tea: drink to your health?" *American Institute for Cancer Research Newsletter,* Spring 1993, Issue 39, p. 5.

Harris, William S., "The prevention of atherosclerosis with antioxidants," *Clinical Cardiology,* 1992, Vol. 15, pp. 636–640.

Hegenauer, Jack, "U.S. RDA vs. RDI: the alphabet soup of nutrition labeling," *Nutrition & the M.D.,* May 1993, p. 5.

Jialal, Iswarlal, and Grundy, Scott M., "Preservation of the endogenous antioxidants in low density lipoprotein by ascorbate but not probucol during oxidative modification," *Journal of Clinical Investigation,* 1991, Vol. 87, pp. 597–601.

Karmas, E., and Harris, R., eds., *Nutritional Evaluation of Food Processing,* 3rd edition. New York: Van Nostrand Reinhold, 1988.

Knekt, Paul, et al., "Serum vitamin E and risk of cancer among Finnish men during a 10-year follow-up," *American Journal of Epidemiology,* 1988, Vol. 127, pp. 28–41.

Kritchevsky, David, "Antioxidant vitamins in the prevention of cardiovascular disease," *Nutrition Today,* January/February 1992, p. 30ff.

"The latest elixir of life: vitamin C," *Tufts University Diet & Nutrition Letter,* July 1992, Vol. 10, No. 5, p. 1ff.

"The latest on links to heart disease," *The Diet-Heart Newsletter,* Vol. 6, No. 2, p. 1ff.

Lehman, J., et al., "Vitamin E in foods form high and low linoleic acid diets," *Journal of the American Dietetic Association,* Sept. 1986, Vol. 86, pp. 1208–1216.

Liebman, Bonnie, "Antioxidants and cancer," *Nutrition Action Health Letter,* July/August 1992, p. 1ff.

Machlin, Lawrence J., "Nutrients as in vivo antioxidants: their role in maintenance of health," *Nutrition & the M.D.,* January 1991, Vol. 17, No. 1, p. 1ff.

Prasad, Kedar N., et al., "Vitamin E and cancer prevention:

recent advances and future potentials," *Journal of the American College of Nutrition*, 1992, Vol. 11, No. 5, pp. 487–500.

"Provisional table on percent retention of nutrients in food preparation," United States Department of Agriculture Human Nutrition Information Service.

"Pumping immunity," *Nutrition Action Healthletter*, April 1993, pp. 5–7.

"Role of antioxidants in heart disease," *Nutrition & the M.D.*, May 1992, Vol. 18, No. 5, p. 1ff.

"Smoking doubles risk of stroke, study finds," *The New York Times*, Nov. 9, 1993, p. C5.

Traber, Maret G., et al., "Vitamin E is delivered to cell via the high affinity receptor for low-density lipoprotein," *American Journal of Clinical Nutrition*, 1984, Vol. 40, pp. 747–751.

"Trendy supplements—diet therapy/obesity update," *Nutrition & the M.D.*, March 1993, Vol. 19, No. 3, p. 8.

"Trio of vitamins are recruited in the fight against heart disease," *Environmental Nutrition*, September 1992, Vol. 15, No. 9, p. 1ff.

"Vitamins C & E wage war against atherosclerosis," *Nutrition & Health News*, Center for Human Nutrition, The University of Texas Southwestern Medical Center at Dallas, Fall 1990, Vol. VII, No. 13, p. 1ff.

Waslien, Carol I., et al., "Micronutrients and antioxidants in processed foods—analysis of data from 1987 food additives survey," *Nutrition Today*, July/August, 1990, p. 36ff.

Watt, Bernice K., and Merrill, Annabel L., *Composition of Foods, Agriculture Handbook No. 8*, Agricultural Research Service, United States Department of Agriculture, 1963, 1975.

Chapter Eight: Toward a Life Free of Free Radicals

Blumberg, Jeffrey B., "Dietary antioxidants and aging," *Contemporary Nutrition*, 1992, Vol. 17, No. 3, p. 1.

Jenkins, R.R., et al., "Introduction: oxidant stress, aging, and exercise," *Medicine and Science in Sports and Exercise*, 1993, Vol. 25, No. 2., pp. 210–212.

Ji, Li Li, "Antioxidant enzyme response to exercise and aging,"

Medicine and Science in Sports and Exercise, 1993, Vol. 25, No. 2, pp. 225–231.

Matsuo, Mitsuyoshi, et al., "Age-related alterations in antioxidant capacity and lipid peroxidation in brain, liver, and lung homogenates of normal and vitamin E-deficient rats," *Mechanisms of Aging and Development,* 1992, Vol. 64, pp. 273–292.

Orr, William C., et al., "The effects of catalase gene overexpression on life span and resistance to oxidative stress in transgenic Drosophilia melanogaster," *Archives of Biochemistry and Biophysics,* August 15, 1992, Vol. 297, No. 1, pp. 35–41.

Paolisso, Guiseppe, et al., "Evidence for a relationship between free radicals and insulin action in the elderly," *Metabolism,* May 1993, Vol. 42, No. 5, pp. 659–663.

Sohal, R.S., "The free radical hypothesis of aging: an appraisal of the current status," *Aging Clinical Exp. Res.,* 1993, Vol. 5, No. 1, pp. 3–17.

Index